Cover photo credits: Clockwise from top left – Dave Marci, file photo, file photo, Blair E. Witherington. Background – file photo.

# Deepwater Horizon Oil Spill Phase II Early Restoration Plan and Environmental Review

Prepared by the Deepwater Horizon Natural Resource Trustees from

State of Alabama (Department of Conservation and Natural Resources; Geological Survey of Alabama)

State of Florida (Department of Environmental Protection; Fish and Wildlife Conservation Commission)

State of Louisiana (Coastal Protection and Restoration Authority; Department of Environmental Quality; Department of Wildlife and Fisheries; Department of Natural Resources; Oil Spill Coordinator's Office)

State of Mississippi (Department of Environmental Quality)

State of Texas (Texas Commission on Environmental Quality; Texas General Land Office; Texas Parks and Wildlife Department)

United States Department of the Interior

National Oceanic and Atmospheric Administration

United States Department of Agriculture

United States Environmental Protection Agency

December 2012

(This page intentionally left blank.)

# EXECUTIVE SUMMARY

## Introduction

The Gulf of Mexico (Gulf) is a priceless national treasure. Its natural resources – water, fish, beaches, reefs, marshes, oil and gas – are the economic engine of the region. The Gulf is likewise vitally important to the entire nation as a bountiful source of food, energy and recreation. The Gulf Coast's unique culture and natural beauty are world-renowned. There is no place like it anywhere else on Earth.

On April 20, 2010 the eyes of the world focused on an oil platform in the Gulf, approximately 50 miles off the Louisiana coast. The mobile drilling unit *Deepwater Horizon*, which was being used to drill an exploratory well for BP Exploration and Production, Inc. (BP), violently exploded, caught fire and eventually sank, tragically killing 11 workers. But that was only the beginning of the disaster. Oil and other substances from the well head immediately began flowing unabated approximately one mile below the surface. Initial efforts to cap the well were unsuccessful, and for 87 days oil spewed unabated into the Gulf. Oil eventually covered a vast area of thousands of square miles, and carried by the tides and currents reached the coast, polluting beaches, bays, estuaries and marshes from the Florida panhandle to west of Galveston Island, Texas. At the height of the spill, approximately 37% of the open water in the Gulf was closed to fishing. Before the well was finally capped, an estimated 5 million barrels (210 million gallons) of oil escaped from the well over a period of approximately 3 months. In addition, approximately 1.84 million gallons of dispersants were applied to the waters of the spill area, both on the surface and at the well head one mile below. Shoreline communities and other responders along the Gulf coast raced to protect coastal habitats as beaches, coastal waters, estuaries, and marshes were put at risk of oiling. Floating booms were placed across inlets, within estuaries, and along sandy beaches creating a barrier to people and to important wildlife habitats. Heavy equipment and lines of workers moved large amounts of sand to form additional berms and barriers. Some response activities to the spill negatively impacted sandy beaches and marshes as thousands of workers descended on the beaches and sensitive wetland areas preparing for the oil to come ashore, searching for oil and removing product by hand and with machines. It was an environmental disaster of unprecedented proportions. It also was a devastating blow to the resource-dependent economy of the region.

While the extent of natural resources impacted by the *Deepwater Horizon* oil spill and response (collectively, "the Spill") is not yet fully evaluated, impacts were widespread and extensive. The full spectrum of the impacts from the Spill, given its magnitude, duration, depth and complexity, will be difficult to determine. The trustees for the Spill, however, are working to assess every aspect of the injury, both to individual resources and lost recreational use of them, as well as the cumulative impacts of the Spill. Affected natural resources include ecologically, recreationally, and commercially important species and their habitats across a wide swath of the coastal areas of Alabama, Florida, Louisiana, Mississippi, and Texas, and a huge area of open water in the Gulf. When injuries to migratory species such as birds, whales, tuna and turtles are considered, the impacts of the Spill could be felt across the United States and around the globe.

## The Role of the Trustees

Under the Oil Pollution Act (OPA), which became law after the 1989 Exxon Valdez oil spill, the federal government, impacted state governments, federally recognized Indian tribes and foreign governments act as "trustees" on behalf of the general public. Trustees are charged with recovering damages from the parties responsible for oil spills and to restore injuries to the public's natural resources. Trustees assess the nature and extent of natural resource injury and develop and implement a restoration plan that involves rehabilitation, replacement, or acquisition of the equivalent of the injured natural resources and services those resources provide under their trusteeship. The *Deepwater Horizon* Trustees (Trustees) are:

- The United States Department of the Interior (DOI), as represented by the National Park Service, United States Fish and Wildlife Service, and Bureau of Land Management;
- The National Oceanic and Atmospheric Administration (NOAA), on behalf of the United States Department of Commerce;
- The United States Department of Agriculture;
- The United States Environmental Protection Agency;
- The State of Louisiana's Coastal Protection and Restoration Authority, Oil Spill Coordinator's Office, Department of Environmental Quality, Department of Wildlife and Fisheries and Department of Natural Resources;
- The State of Mississippi's Department of Environmental Quality;
- The State of Alabama's Department of Conservation and Natural Resources and Geological Survey of Alabama;
- The State of Florida's Department of Environmental Protection and Fish and Wildlife Conservation Commission;
- And for the State of Texas: Texas Parks and Wildlife Department, Texas General Land Office and Texas Commission on Environmental Quality.[1]

The Trustees began working together in the early days of the Spill. The result has been an unprecedented state-federal collaboration, with a unity of vision and purpose, and a strong desire by all the Trustees to act as quickly as possible to restore the Gulf. Trustee efforts to assess the injuries to natural resources began within hours of the explosion and continue to the present. The Trustees uniformly believe that restoration of the natural resources in the Gulf must begin as soon as possible. This Phase II Early Restoration Plan and Environmental Review (Phase II ERP/ER) contains the plan for the second set of restoration actions that will be undertaken by the Trustees, paid for by those responsible for injuries to natural resources and the services they provide, representing a step on the road to a full recovery for the Gulf. The ultimate goal of the Trustees is comprehensive and long lasting repairs to the Gulf ecosystem, and the communities that depend on it, to the condition they would have been in if the Spill had not occurred (i.e., the baseline conditions), as well as to compensate the public for its lost use of the resources during the time they were injured.

From the outset, the Trustees expected that the restoration of resources injured by the Spill would be a massive undertaking, and that during the assessment, injuries would continue to accrue. The

---

[1] The Department of Defense (DOD) is also a trustee of natural resources associated with DOD-managed land on the Gulf Coast, which is included in the ongoing natural resource damage assessment (NRDA).

Trustees decided that because of the pervasive and ongoing nature of the damages to natural resources in the region, it would be in the best interest of the public to accelerate restoration and begin implementing projects, if possible, even before completion of the full damage assessment. The Trustees approached BP in the fall of 2010, and negotiations on an early restoration fund commenced.

Exactly one year after the explosion on the *Deepwater Horizon* rig, the Trustees and BP entered into an unprecedented agreement whereby BP set aside one billion dollars to fund early restoration projects agreed to by BP and the Trustees, incorporating public review. This early restoration agreement, known as the "Framework Agreement,"[2] represents the initial step toward the restoration of natural resources injured by the *Deepwater Horizon* Spill. It is a down payment against the ultimate claim for damages from the Spill. The Trustees expect, pending agreement with BP, to be able to fund more early restoration projects in addition to the eight projects addressed in the Phase I Early Restoration Plan and Environmental Assessment (Phase I ERP/EA; Trustees, 2012) and the two projects selected herein. The Trustees continue to assess the injuries to natural resources and services resulting from the Spill and pursue the ultimate claim for damages. Restoration work will take many years to complete, and long-term monitoring and adaptive management of the Gulf ecosystem will likely continue for decades until the Trustees can be certain that the public has been fully compensated for its losses.

**Early Restoration Project Selection**

Following signature of the Framework Agreement, the Trustees invited the public to provide early restoration project ideas and proposals. The Trustees received hundreds of proposals, which were made publicly available at http://www.gulfspillrestoration.noaa.gov/restoration/give-usyour-ideas/view-submitted-projects/. The Trustees implemented a project selection process to evaluate proposals and ensure that restoration would begin as soon as possible. Figure ES-1 depicts the general selection process, which included project solicitation, project screening and identification, negotiation, public review and comment, and final selection.

The Trustees evaluated potential early restoration projects using criteria included in applicable damage assessment and restoration regulations and programs, the Framework Agreement, and factors that are otherwise key components in planning early restoration. Under OPA regulations, restoration alternatives are evaluated with regard to:

- The cost to carry out the alternative;
- The extent to which each alternative is expected to meet the Trustees' goals and objectives in returning the injured natural resources and services to baseline and/or compensating for interim losses (the ability of the restoration project to provide comparable resources and services, that is, the nexus between the project and the injury);
- The likelihood of success of each alternative;
- The extent to which each alternative will prevent future injury as a result of the incident, and avoid collateral injury as a result of implementing the alternative;

---

[2] See http://www.restorethegulf.gov/sites/default/files/documents/pdf/framework-for-early-restoration-04212011.pdf.

- The extent to which each alternative benefits more than one natural resource and/or service; and
- The effect of each alternative on public health and safety.

Under OPA regulations, if the Trustees conclude that two or more restoration alternatives are equally preferable, the most cost-effective alternative must be chosen.

In addition, the Framework Agreement provides that early restoration projects meet the following criteria:

- Contribute to making the environment and the public whole by restoring, rehabilitating, replacing, or acquiring the equivalent of natural resources or services injured as a result of the Spill, or compensating for interim losses resulting from the incident;
- Address one or more specific injuries to natural resources or services associated with the incident;
- Seek to restore natural resources, habitats, or natural resource services of the same type, quality, and of comparable ecological and/or human-use value to compensate for identified resource and service losses resulting from the incident;
- Are not inconsistent with the anticipated long-term restoration needs and anticipated final restoration plan; and
- Are feasible and cost-effective.

In early restoration planning, the Trustees are also taking into account several practical considerations that, while not legally mandated, are nonetheless useful and permissible to help screen the large number of potential qualifying projects. None of these practical considerations are used as a "litmus test"; rather, they are used as flexible, discretionary factors to supplement the decision criteria described above. For example, Trustees:

- Take into account how quickly a given project is likely to begin producing environmental benefits;
- Seek a diverse set of projects providing benefits to a broad array of potentially injured resources;
- Focus on types of projects with which they have significant experience, allowing them to predict costs and likely success with a relatively high degree of confidence and making it easier to reach agreement with BP on the Offsets (see Section 1.3) attributed to each project, as required by the Framework Agreement; and
- Give preference to projects that were closer to being ready to implement.

The Trustees acted promptly in 2011 to identify project proposals that met selection criteria, and then narrowed the potential project list down to an initial group to move forward into discussion with BP on cost and Offsets. The Trustees and BP came to preliminary agreement on a set of proposals, which the Trustees proposed as Phase I projects in a Draft Phase I ERP/EA released for public comment in December 2011 and finalized as the "*Deepwater Horizon* Oil Spill Phase I Early Restoration Plan and Environmental Assessment" in April 2012 (Trustees, 2012).

Partially in response to some specific public comments received on the Phase I Draft Early Restoration Plan (DERP)/EA, the Trustees proposed two more early restoration projects to address injuries to the nesting habitat of beach nesting birds and of nesting loggerhead sea turtles that resulted from response activities to the Spill. These two projects were included in the Draft Phase II ERP/ER released for public comment on November 6, 2012. These projects were proposed at this time because loggerhead sea turtles and beach nesting birds begin nesting along the Northeast Gulf coast in February and implementation of these projects needs to begin in advance of nesting season to provide benefits during the 2013 nesting season. A public meeting was also held on November 13, 2012 in Pensacola, Florida to facilitate public review and comment. The Trustees accepted comment on the proposed plan through December 10, 2012.

## Selected Projects

Consistent with OPA and the National Environmental Policy Act, the Trustees considered public comment prior to final selection of these Phase II projects. A summary of comments on the Draft Phase II ERP/ER, the Trustees' responses to comments and the final selected Phase II projects are included in this final "*Deepwater Horizon* Oil Spill Phase II Early Restoration Plan and Environmental Review" (Phase II ERP/ER), together with the Trustees' environmental review documentation. In addition, this Phase II ERP/ER includes a description and quantification of the Offsets preliminarily agreed to by BP and the Trustees.

This Phase II ERP/ER consists of the two projects listed in Table ES-1, and more fully described in this document. They address response injuries to habitat of beach nesting birds and of nesting loggerhead sea turtles and have project components located in Florida, Alabama and Mississippi. While this plan includes two projects, each project was viewed and evaluated as independent from the other.

It is important to emphasize that restoration proposals developed pursuant to the Framework Agreement are not intended to provide the full extent of restoration needed to satisfy the Trustees' claims against BP. Restoration will continue until the public is fully compensated for the natural resources and services that were lost as a result of the Spill.

## Next Steps

This Phase II ERP/ER serves as the Trustees' final selection of Phase II early restoration projects, taking into account the suite of potential projects proposed, the NRDA and Framework Agreement process, and public comment on the Draft Phase I ERP/EA and Draft Phase II ERP/ER. Per the Framework Agreement, the Trustees will move forward with agreements with BP to fund these projects and commence implementation, as described in more detail throughout this document. Updates on the progress of project implementation will be available at http://www.gulfspillrestoration.noaa.gov.

As previously noted, the projects selected in this Phase II ERP/ER represent only the second set of projects in the early restoration process. The Trustees continue to evaluate projects already submitted by the public for consideration, as well as any new projects as they are received, with the intent of proposing additional projects until funds made available under the Framework Agreement are exhausted. It is important to emphasize that restoration proposals developed pursuant to the Framework Agreement are not intended to provide the full extent of restoration needed to satisfy the Trustees' claims against BP. At the end of the NRDA process, the Trustees will credit all the Offsets identified for approved early restoration projects against their assessment of the **total** injury for the Spill. Restoration beyond early restoration projects will be required to fully compensate the public for natural resource losses from the Spill and will continue until the public is fully compensated for the natural resources and services that were lost as a result of the Spill.

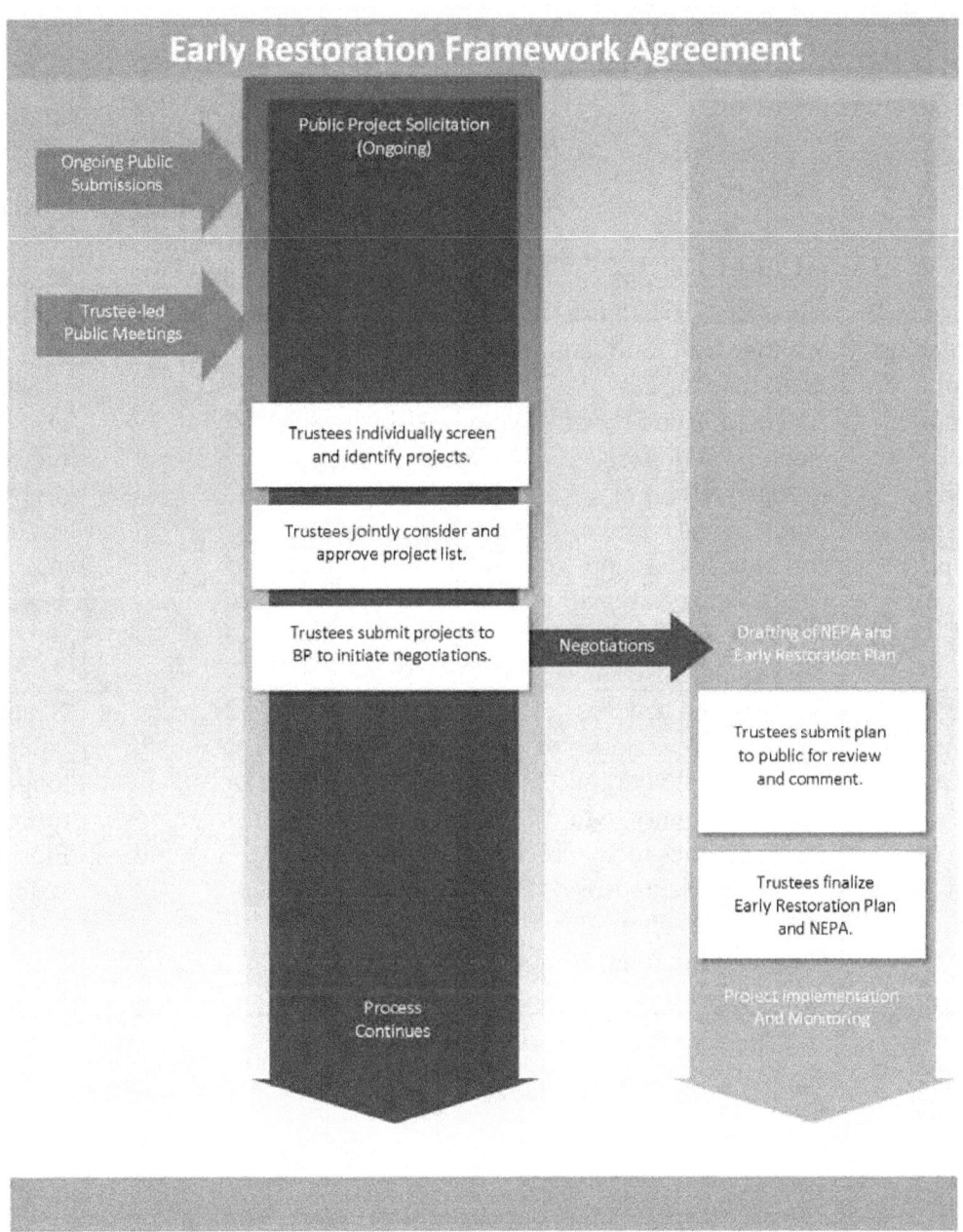

Figure ES-1. General early restoration project selection process.

Table ES-1. Early restoration projects included in the selected Alternative.

| Project Title | Location | Selected Restoration | Estimated Cost (including potential contingencies)[3] | Resources Benefitted |
|---|---|---|---|---|
| Enhanced Management of Avian Breeding Habitat Injured by Response in the Florida Panhandle, Alabama, and Mississippi | Florida: Escambia, Santa Rosa, Okaloosa, Walton, Bay, Gulf, and Franklin counties. Alabama: Bon Secour National Wildlife Refuge (NWR) in Baldwin and Mobile counties. Mississippi: Gulf Islands National Seashore (GUIS) – Mississippi District. | Symbolic fencing, predator control, and stewardship around important nesting areas to prevent disturbance | $4,658,118 | Nesting and foraging habitat for beach nesting birds in Florida, and on DOI lands in Alabama and Mississippi. |
| Improving Habitat Injured by Spill Response: Restoring the Night Sky | State-owned beaches within the boundaries of the Gulf State Park in Baldwin County, AL, and properties in Escambia, Santa Rosa, Okaloosa, Walton, Bay, Gulf, and Franklin counties, FL. | Reduce artificial lighting impacts on nesting habitat for loggerhead sea turtles | $4,321,165 | Nesting habitat for loggerhead sea turtles in Florida and state lands in Alabama. |

---

[3] Actual costs may differ depending on future contingencies, but will not exceed the amount shown without further agreement between the Trustees and BP.

# TABLE OF CONTENTS

# CHAPTER 1  BACKGROUND, PURPOSE AND NEED FOR PROPOSED ACTION

## 1.1  Introduction

On or about April 20, 2010, the mobile offshore drilling unit *Deepwater Horizon,* which was being used to drill a well for BP Exploration and Production, Inc. (BP) in the Macondo prospect (Mississippi Canyon 252 – MC252), experienced an explosion, leading to a fire and its subsequent sinking in the Gulf of Mexico (the Gulf). This incident resulted in discharges of oil and other substances from the rig and the submerged wellhead into the Gulf. An estimated 5 million barrels (210 million gallons) of oil were subsequently released from the well over a period of approximately 3 months.[4] In addition, approximately 1.84 million gallons of dispersants[5] were applied to the waters of the spill area in an attempt to minimize impacts from spilled oil (National Commission on the BP Deepwater Horizon Oil Spill and Offshore Drilling, 2011).

The U.S. Coast Guard responded and directed federal efforts to contain and clean up the spill (hereafter referred to as "the Spill," which includes activities conducted in response to the spilled oil). At one point nearly 50,000 responders were involved in cleanup activities in open water, beach and marsh habitats. The magnitude of the Spill was unprecedented, causing impacts to coastal and oceanic ecosystems ranging from the deep ocean floor, through the oceanic water column, to the highly productive coastal habitats of the northern Gulf, including estuaries, shorelines and coastal marsh. Affected resources include ecologically, recreationally, and commercially important species and their habitats in the Gulf and along the coastal areas of Alabama, Florida, Louisiana, Mississippi, and Texas. These fish and wildlife species and their supporting habitats provide a number of important ecological and human use services.

This Phase II Early Restoration Plan and Environmental Review (ERP/ER) includes the second set of early restoration projects being selected by the *Deepwater Horizon* Trustees (Trustees) to address natural resource injuries resulting from the Spill. The two selected projects address response injuries to nesting habitat for beach nesting birds and nesting loggerhead sea turtles. Because loggerhead sea turtles and beach nesting birds begin nesting along the Northeast Gulf coast in February, the Trustees recognized these projects needed to be implemented in a timely manner to be effective during the 2013 nesting season, and therefore proposed while additional early restoration projects are being developed in accordance with the Framework Agreement (see Section 1.8).

---

[4] Oil Budget Team, OIL BUDGET CALCULATOR TECHNICAL DOCUMENTATION (November 23, 2010).
[5] Dispersants do not remove oil from the ocean. Rather, they are used to help break large globs of oil into smaller droplets that can be more readily dissolved into the water column.

## 1.2 Overview of the Oil Pollution Act and the National Environmental Policy Act

### 1.2.1 The Oil Pollution Act

The Oil Pollution Act (OPA) Title 33 United States Code (U.S.C.) § 2701. *et seq.*, and the regulations for natural resource damage assessments (NRDAs) under OPA, 15 Code of Federal Regulations (C.F.R.) Part 990, establish a liability regime for oil spills into navigable waters or adjacent shorelines that injure or are likely to injure natural resources and services that those resources provide to the ecosystem or humans. Pursuant to section 1006 of OPA, federal and state trustees for natural resources are authorized to (1) assess natural resource injuries resulting from a discharge of oil or the substantial threat of a discharge and response activities, and (2) develop and implement a plan for restoration of such injured resources.

The federal trustees are designated pursuant to section 1006(b)(2) of OPA and Executive Orders 12777 and 13626. The following federal agencies are designated natural resources trustees under OPA and are currently acting as trustees for the Spill[6]:

- The United States Department of the Interior (DOI), as represented by the National Park Service (NPS), United States Fish and Wildlife Service (FWS), and Bureau of Land Management;
- The National Oceanic and Atmospheric Administration (NOAA), on behalf of the United States Department of Commerce;
- The United States Department of Agriculture (USDA); and
- The United States Environmental Protection Agency (EPA).

State trustees are designated by the Governors of each state pursuant to section 1006(b)(3) of OPA and Executive Orders 12777 and 13626. The following state agencies are designated natural resources trustees under OPA and are currently acting as trustees for the Spill:

- The State of Louisiana's Coastal Protection and Restoration Authority, Oil Spill Coordinator's Office, Department of Environmental Quality, Department of Wildlife and Fisheries and Department of Natural Resources;
- The State of Mississippi's Department of Environmental Quality;
- The State of Alabama's Department of Conservation and Natural Resources and Geological Survey of Alabama;
- The State of Florida's Department of Environmental Protection (FDEP) and Fish and Wildlife Conservation Commission (FWC); and
- For the State of Texas: Texas Parks and Wildlife Department, Texas General Land Office and Texas Commission on Environmental Quality.

---

[6] The Department of Defense (DOD) is also a trustee of natural resources associated with DOD-managed land on the Gulf Coast, which is included in the ongoing NRDA.

In addition to acting as Trustees for this incident under OPA, the States of Louisiana, Mississippi, Alabama, Florida and Texas are also acting pursuant to their applicable state laws and authorities, including:

- The Louisiana Oil Spill Prevention and Response Act of 1991, La. R.S. 30:2451 *et seq.*, and accompanying regulations, La. Admin. Code 43:101 *et seq.*;
- The Texas Oil Spill Prevention and Response Act, Tex. Nat. Res. Code, Chapter 40.01 *et seq.*;
- The Florida Pollutant Discharge Prevention and Removal Act, Fla. Statutes Section 376.011 *et seq.*;
- The Mississippi Air and Water Pollution Control Law, Miss. Code Ann. §§ 49-17-1 through 49-17-43; and
- Alabama Code §§ 9-2-1 *et seq.* and 9-4-1 *et seq.*

Pursuant to OPA, federal and state agencies, Indian tribes and foreign governments may act as trustees on behalf of the public to assess the injuries and plan for restoration to compensate for those injuries. OPA further instructs the designated trustees to develop and implement a plan for the restoration, rehabilitation, replacement, or acquisition of the equivalent of the injured natural resources under their trusteeship (hereafter collectively referred to as "restoration"). OPA defines "natural resources" to include land, fish, wildlife, biota, air, water sources, and other such resources belonging to, managed by, held in trust by, appertaining to, or otherwise controlled by the United States, any State or local government or Indian tribe, or any foreign government. This Phase II ERP/ER was prepared jointly by the Trustees.

Natural resource services are the ecological and human use services that natural resources provide. Examples of ecological services include biological diversity, nutrient cycling, food production for other species, habitat provision, and other services that natural resources provide for each other. Human use services include activities that make 'direct' use of natural resources (e.g., boating, nature photography, education, fishing, swimming, hiking, etc.) as well as the value the public holds for natural resources independent of their own use of such resources (e.g., existence value, bequest value, etc.). For the purposes of this document the term "natural resource services" shall include these ecological and human use services.

### 1.2.2 The National Environmental Policy Act

The National Environmental Policy Act (NEPA), 42 U.S.C. § 4321, *et seq.* and its implementing regulations at 40 C.F.R. Parts 1500-1508 set forth a process of impact analysis and public review for federal agency actions, including restoration actions. NEPA provides a mandate and a framework for federal agencies to consider all reasonably foreseeable environmental effects of their proposed actions and to inform and involve the public in their environmental analysis and decision-making process.

Actions undertaken by federal trustees to restore natural resources or services under OPA and other federal laws are subject to NEPA, 42 U.S.C. § 4321 *et seq.*, and the regulations guiding its implementation at 40 C.F.R. Part 1500.[7] NEPA and its implementing regulations outline the

---

[7] NEPA imposes legal requirements on federal trustees only.

5

responsibilities of federal agencies under NEPA, including the preparation of environmental documentation. In general, federal agencies contemplating implementation of a major federal action must produce an environmental impact statement (EIS) if the action is expected to have significant impacts on the quality of the human environment. When it is uncertain whether a contemplated action is likely to have significant impacts, federal agencies prepare an environmental assessment (EA) to evaluate the need for an EIS. If the EA demonstrates that the proposed action will not significantly impact the quality of the human environment, the federal agencies issue a Finding of No Significant Impact (FONSI), which satisfies the requirements of NEPA, and no EIS is required. If a FONSI cannot be made, then an EIS is required. Additionally, over time, through study and experience, agencies may identify activities that do not need to undergo detailed environmental analysis in an EA or an EIS because the activities do not individually or cumulatively have a significant effect on the human environment. Agencies can define categories of such activities, called categorical exclusions (CXs), in their NEPA implementing procedures, as a way to reduce unnecessary paperwork and delay.

The Trustees prepared this Phase II ERP/ER in accordance with OPA NRDA regulations (see 15 C.F.R § 990.23) and NEPA requirements, which both require public involvement in the decision-making process. This Phase II ERP/ER presents information to the public regarding the affected environment, NRDA restoration planning, and actions designed to help address natural resource injuries and lost human use of injured natural resources caused by the Spill. Restoration projects go beyond cleanup activities by restoring[8] injured natural resources or lost services.

The Phase II restoration alternative selected by the Trustees (see Chapter 3) is comprised of two restoration projects. As discussed in Chapter 4, each project has been evaluated separately under NEPA because each project has independent utility. In accordance with NEPA and its implementing regulations, this Phase II ERP/ER summarizes the current environmental setting, describes the purpose and need for restoration, identifies restoration alternatives considered for injuries, assesses their applicability and potential environmental consequences, and summarizes the opportunity afforded for public participation in the process of making the Phase II early restoration plan decisions. This information has been used to make a threshold determination as to whether preparation of an EIS is required prior to selecting the final Phase II early restoration actions.

### 1.2.3 Compliance with Other Applicable Authorities

In addition to the requirements of OPA and NEPA, requirements of other laws may apply to the early restoration planning or early restoration implementation. The Trustees will ensure compliance with all applicable authorities for all early restoration projects. To assist the public with identifying other applicable authorities, the Trustees prepared a non-exclusive list of other potentially applicable federal authorities, attached as Appendix B. Whether and the extent to which an authority applies to a particular project depends on the specific characteristics of a particular project. Consequently, not every authority listed in Appendix B would apply to every project. In addition, state Trustees will ensure compliance with applicable authorities in their individual states.

---

[8] For the purposes of this document, "restoring" or "restoration" includes any action that restores, rehabilitates, replaces, or acquires the equivalent of the injured natural resources or lost services.

## 1.3    Natural Resource Damage Assessment Restoration Planning

Restoration activities are intended to restore or replace habitats, species, and services to their baseline condition, defined as the condition of the natural resources and services that would have existed had the incident not occurred (primary restoration), and to compensate the public for interim losses from the time natural resources are injured until they are restored or replaced to achieve baseline conditions (compensatory restoration). To meet these goals, the restoration activities need to produce benefits that are related, or have a nexus, to natural resources injured and associated service losses resulting from the Spill and associated response or clean-up activities.

> **Restoration Terms Defined**
>
> Restoration: Any action that restores, rehabilitates, replaces, or acquires the equivalent of the injured natural resources.
>
> Primary Restoration: Any action that replaces or restores injured natural resources and services to their baseline condition.
>
> Compensatory Restoration: Any action that replaces or restores the natural resource injuries and services lost from the date of injury until recovery to baseline conditions occurs.

NRDA restoration planning is designed to evaluate potential injuries to natural resources and natural resource services; to use that information to determine whether and to what extent restoration is needed; to identify potential restoration actions to address that need; and to provide the public with an opportunity to review and comment on the proposed restoration alternatives. Restoration planning has two basic components: (1) injury assessment and (2) restoration selection.

The goal of injury assessment is to determine the nature and extent of injuries to natural resources and services. The goal of restoration planning is to evaluate the need for and type of restoration required based on the injury assessment. Ultimately, Trustees identify proposed restoration alternatives expected to compensate the public for losses of natural resources and services resulting from the Spill.

Given its expansive geographic scale and complexity, the *Deepwater Horizon* NRDA may continue for years. In response to this extraordinary event, the Trustees initiated the restoration and planning efforts described below, even while damage assessment activities continue. The early restoration projects selected in this Phase II ERP/ER are not intended to fully compensate the public for injuries caused by the Spill. Additional restoration actions will be required.

### Emergency Restoration

Under OPA, trustees may take emergency restoration actions before completing the NRDA process in order to minimize continuing, or prevent additional, injury as long as the actions are feasible and the cost of the actions are reasonable.

The Trustees collectively implemented three emergency restoration projects as part of the Spill, addressing submerged aquatic vegetation (SAV), waterfowl, and sea turtles. The SAV project was implemented to prevent additional injury by restoring SAV beds damaged by propeller scarring and other response vessel impacts. The waterfowl habitat enhancements project provided alternative wetland habitat in Mississippi for waterfowl and shorebirds that might otherwise winter in Spill-affected habitats. The sea turtle project was completed to improve the nesting and hatching success of endangered sea turtles on the Texas coast, including Padre Island National Seashore, during the Spill. Some Trustees also implemented additional response and emergency restoration actions independent of the other Trustees.

**Gulf Spill Restoration Planning Programmatic Environmental Impact Statement**

The Trustees are preparing a draft programmatic environmental impact statement (DPEIS) to address environmental impacts from and to facilitate the development of a draft programmatic restoration plan. Public input from scoping conducted as part of that process, and similar exercises conducted by individual Trustees, will also be considered in the development of early restoration plans (see Section 1.5 below). The DPEIS will assist the Trustees in making informed decisions regarding the selection and implementation of a range of restoration types that could be used to compensate the public and the environment for the loss of natural resources and services from the Spill. The Notice of Intent initiating this effort can be viewed at: http://www.gulfspillrestoration.noaa.gov/wp-content/uploads/2011/02/PEIS-NOI_signed.pdf.

**Early Restoration**

On April 20, 2011, the Trustees entered into an agreement whereby BP is to provide $1 billion toward early restoration projects in the Gulf to address injuries to natural resources caused by the Spill. As described below, this early restoration agreement, entitled "Framework for Early Restoration Addressing Injuries Resulting from the *Deepwater Horizon* Oil Spill" (Framework Agreement),[9] represents a preliminary, initial step toward the restoration of natural resources injured by the Spill. The Framework Agreement is intended to facilitate and expedite restoration in the Gulf in advance of the completion of the NRDA process. The Framework Agreement provides a mechanism through which the Trustees and BP can work together "to commence implementation of early restoration projects that will provide meaningful benefits to accelerate restoration in the Gulf as quickly as practicable" prior to completion of the NRDA process or full resolution of the Trustees' natural resource damage claims.

This Phase II ERP/ER addresses OPA and NEPA requirements for implementing Phase II early restoration projects. It includes a discussion of the alternatives considered for Phase II as well as the environmental review for each of the selected projects. Early restoration plans are not intended to quantify the extent of restoration needed to satisfy claims under applicable law against the responsible parties; rather, the early restoration projects described herein are intended to expedite the overall restoration process.

The Phase II ERP/ER also identifies the restoration benefits estimated to be provided by each project (referred to as "Offsets"). The term "Offsets" shall have the same meaning as provided in the Framework Agreement. Pursuant to the Framework Agreement, the Offsets were estimated using metrics that reflect natural resources and/or services expected to result from each project. At the end of the NRDA process, the Trustees will credit the Offsets identified for these early

---

[9] http://www.restorethegulf.gov/sites/default/files/documents/pdf/framework-for-early-restoration-04212011.pdf.

restoration projects against their assessment of the total injury for the Spill. Further restoration will still be required to fully compensate the public for natural resource losses from the Spill.

The Draft Phase II ERP/ER included an evaluation of a No Action alternative (Alternative A) and an evaluation of the two proposed early restoration projects (Alternative B). Under Alternative A (No Action – Natural Recovery), the Trustees would not implement any additional early restoration projects at this time. Selecting this alternative would not preclude analysis and implementation of additional restoration activities at a later date. The selected alternative (Alternative B: Phase II Early Restoration Projects) describes two projects that the Trustees concluded, after considering public comment on the Draft Phase II ERP/ER, meet the evaluation criteria described in more detail in Section 1.6, and require implementation in a timely fashion so that habitats are improved in time for the 2013 Gulf nesting season for beach nesting birds and loggerhead sea turtles. It is important to note that the proposed projects in this Phase II ERP/ER represent only a small portion of the early restoration projects being considered by the Trustees. The Trustees will continue to evaluate projects already submitted for consideration – as well as any new projects as they are received – with the intent of proposing additional projects.

In pursuing these projects and other early restoration options, the Trustees are also mindful of other Gulf restoration reports and related efforts, such as those by the Gulf Coast Ecosystem Restoration Task Force (GCERTF, 2011), Mabus (2010), Brown et al. (2011), Peterson et al. (2011), Resources and Ecosystems Sustainability, Tourist Opportunities, and Revived Economies of the Gulf Coast States Act of 2012 (RESTORE Act) (title I, subtitle F of Public Law 112-141) and others, including restoration planning efforts being undertaken by individual Trustees, such as Louisiana's Coastal Master Plan and Annual Plan updates, the Mississippi Coastal Improvements Plan (USACE, 2009) and NRCS (2011).

## 1.4    Purpose and Need for Early Restoration

The Phase II early restoration projects selected in this plan are designed to accelerate meaningful restoration in the Gulf and compensate the public for injuries to beach nesting habitats prior to completion of the full damage assessment. The projects are not intended to, and do not fully, address all injuries caused by the Spill.

## 1.5    Restoration Project Solicitation

Public input is an integral part of NEPA, OPA and the Spill restoration planning effort. Public review allows the public to consider and provide direct input to the Trustees on proposed restoration plans and alternatives and ensures that the Trustees can consider relevant information and concerns of the public prior to making final decisions on proposed actions.

Following the Spill, the Trustees established websites to provide the public information about injury and restoration processes.[10] A Notice of Intent to Conduct Restoration Planning for the *Deepwater Horizon* Oil Spill (Notice) was published in the Federal Register on October 1, 2010 and announced publicly by the Trustees. Pursuant to 15 C.F.R. § 990.44, the Notice announced that the Trustees determined to proceed with restoration planning to fully evaluate, assess, quantify, and develop plans for restoring, replacing, or acquiring the equivalent of natural resources injured and losses resulting from the Spill. Public solicitation of restoration projects has been ongoing since publication of the Notice. The Trustees invited the public to participate in restoration planning for the Spill in accordance with 15 C.F.R. § 990.14(d) and State authorities, including hosting public meetings held across all the Gulf States during October, November and December 2010. A complete record of the public meetings and input opportunities is available at http://www.gulfspillrestoration.noaa.gov.

The Trustees have addressed and continue to address NRDA, the restoration planning process and potential restoration projects at public meetings and venues and meet with many non-governmental organizations and other potential stakeholders. The Trustees continue to solicit restoration ideas via the web[11] and continue to consider existing and new project proposals as part of the restoration planning process. Figure 1 depicts the general project solicitation and selection process for early restoration. In summary, early restoration project selection is a step-wise process comprised of: (1) project solicitation; (2) project screening and identification; (3) negotiation; and (4) public review and comment, described more fully below.

## 1.6    Evaluation Criteria

In evaluating potential early restoration actions, the Trustees consider the broad suite of projects proposed through the project solicitation process. Proposals are evaluated based on criteria included in the OPA NRDA regulations, the Framework Agreement, as well as factors that are otherwise key components in planning or effecting early restoration, including those associated with other laws, regulations and programs. The OPA NRDA regulations (15 C.F.R. § 990.54) provide guidance concerning the evaluation and selection of projects designed to compensate the public for injuries caused by oil spills. These regulations require the Trustees to evaluate proposed restoration alternatives based on, at a minimum:

- The cost to carry out the alternative;
- The extent to which each alternative is expected to meet the Trustees' goals and objectives in returning the injured natural resources and services to baseline and/or compensating for interim losses (the ability of the restoration project to provide comparable resources and services, that is, the nexus between the project and the injury, is an important consideration in the project selection process);
- The likelihood of success of each alternative;

---

[10] See www.fws.gov/contaminants/DeepwaterHorizon/DH_NRDA.cfm; www.gulfspillrestoration.noaa.gov; losco-dwh.com; www.dep.state.fl.us/deepwaterhorizon; www.mdeqnrda.com; http://www.tpwd.state.tx.us/landwater/water/environconcerns/damage_assessment/deep_water_horizon.phtm; www.outdooralabama.com.

[11] See www.gulfspillrestoration.noaa.gov; losco-dwh.com; www.mdeqnrda.com; http://www.tpwd.state.tx.us/landwater/water/environconcerns/damage_assessment/deep_water_horizon.phtml www.outdooralbama.com, www.dep.state.fl.us/deepwaterhorizon.

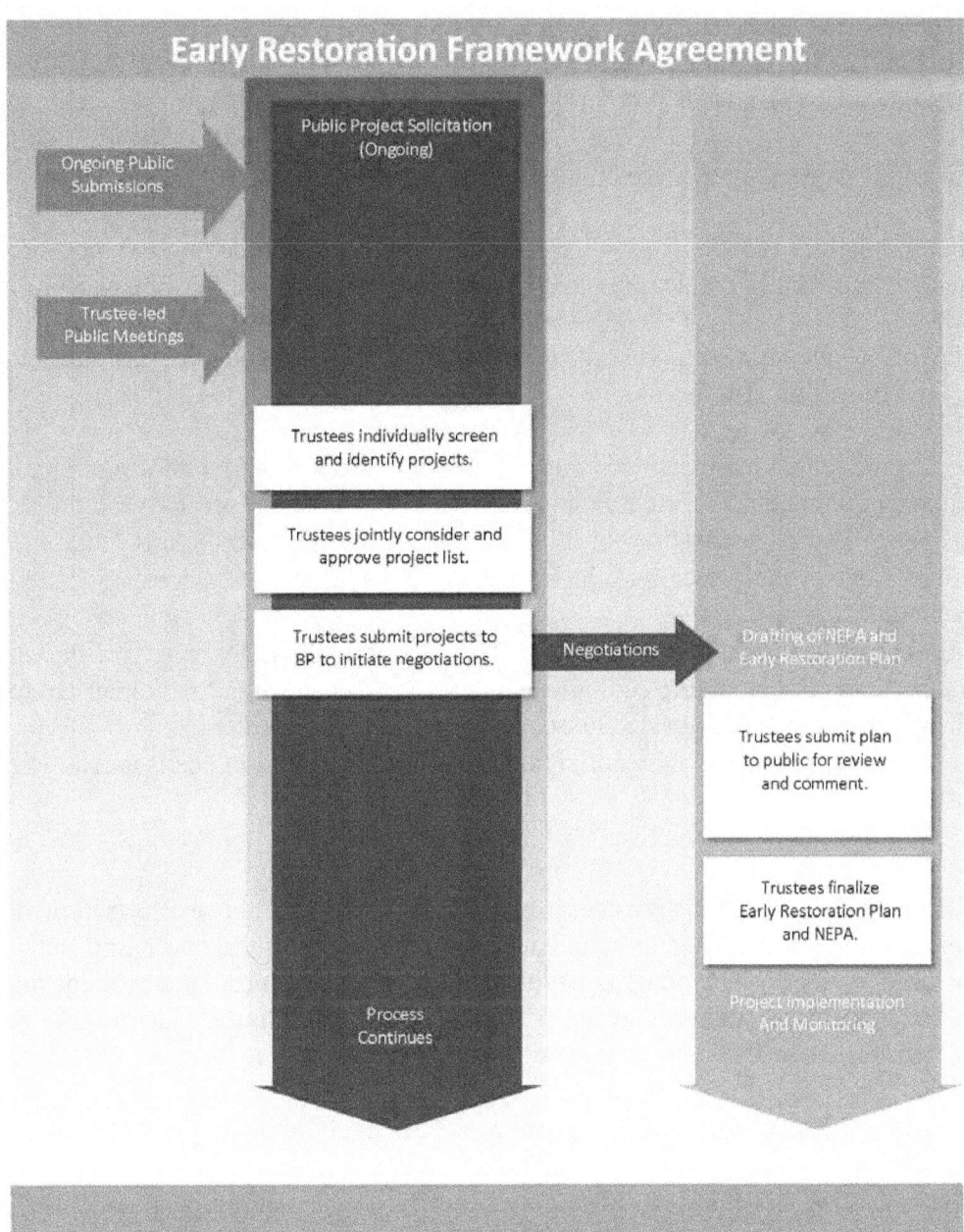

Figure 1. General Early Restoration project selection process.

- The extent to which each alternative will prevent future injury as a result of the incident, and avoid collateral injury as a result of implementing the alternative;
- The extent to which each alternative benefits more than one natural resource and/or service; and
- The effect of each alternative on public health and safety.

Under OPA regulations (15 C.F.R. § 990.54), if the Trustees conclude that two or more alternatives are equally preferable, the most cost-effective alternative must be chosen.

The Framework Agreement states that the Trustees shall select projects for early restoration that meet all of the following criteria:

- Contribute to making the environment and the public whole by restoring, rehabilitating, replacing, or acquiring the equivalent of natural resources or services injured as a result of the Spill, or compensating for interim losses resulting from the incident;
- Address one or more specific injuries to natural resources or services associated with the incident;
- Seek to restore natural resources, habitats, or natural resource services of the same type, quality, and of comparable ecological and/or human-use value to compensate for identified resource and service losses resulting from the incident;
- Are not inconsistent with the anticipated long-term restoration needs and anticipated final restoration plan; and
- Are feasible and cost-effective.

In early restoration planning, the Trustees are also taking into account several practical considerations that, while not legally mandated, are nonetheless useful and permissible to help screen the large number of potential qualifying projects. None of these practical considerations are used as a "litmus test"; rather, they are used as flexible, discretionary factors to supplement the decision criteria described above. For example, Trustees:

- Take into account how quickly a given project is likely to begin producing environmental benefits;
- Seek a diverse set of projects providing benefits to a broad array of potentially injured resources;
- Focus on types of projects with which they have significant experience, allowing them to predict costs and likely success with a relatively high degree of confidence and making it easier to reach agreement with BP on the Offsets attributed to each project, as required by the Framework Agreement; and
- Give preference to projects that are closer to being ready to implement.

All of these discretionary factors are consistent with a key objective for pursuing early restoration: to secure tangible recovery of natural resources and natural resource services for the public's benefit while the longer-term process of fully assessing injury and damages is still underway.

In addition, OPA regulations (15 C.F.R. § 990.56) include specific guidance on the utilization of existing restoration projects and regional restoration plans to address natural resource injuries when appropriate [e.g., Louisiana Regional Restoration Plan, Region 2, NOAA et al., 2007; Louisiana Regional Restoration Planning Program (RRP Program)].[12] Projects already developed under such plans, with engineering designs, cost analyses, partner coordination, and permit and NEPA requirements satisfied, could be implemented quickly, and are good candidates for consideration in the early restoration process.

## 1.7   The Early Restoration Project Selection Process

The process that resulted in the selected alternative presented in this Phase II ERP/ER was developed by the Trustees to be responsive to the purpose and need for conducting early restoration. The Trustees identified the projects selected in this ERP/ER as part of their continuing effort to act promptly to identify project proposals that meet the above criteria. The project selection process for early restoration, as discussed below, is a phased process; multiple rounds of project identification, negotiating, and public comment will continue per the provisions of the Framework Agreement. The Trustees will continue to collect and consider project proposals for subsequent rounds of early restoration.

## 1.8   Project Negotiation with BP

The OPA NRDA regulations require the Trustees to invite responsible parties to participate in the NRDA process. However, the authority and responsibility to assess natural resource injuries and losses and to define appropriate restoration plans rests solely with the Trustees. BP confirmed its interest in cooperatively participating in the NRDA process in 2010. The Framework Agreement evidences BP's willingness to support planning and implementing early restoration.

The process for selecting early restoration projects under the Framework Agreement began with project solicitation, development and evaluation by the Trustees as discussed above. The Framework Agreement requires the Trustees and BP to agree on (1) the funding amount for a proposed project, and (2) Offsets. After the Trustees and BP reached an agreement in principle on these terms for the two projects, these projects were combined into the Trustees' proposed alternative in the Phase II Draft Early Restoration Plan (DERP)/ER. However, the agreements can be finalized only after the public review process, described in more detail below.

## 1.9   Public Review and Comment

OPA, NEPA and the Framework Agreement require public input into the restoration process associated with the Spill. The Phase II DERP/ER served as a proposed restoration plan for early restoration, the environmental review of the projects under NEPA, and the means used by the

---

[12] Louisiana's RRP Program identifies the statewide Program structure, defines those trust resources and services in Louisiana that are likely to be or are anticipated to be injured (*i.e.,* at risk) by oil spill incidents, establishes a decision-making process, and sets forth criteria that are used to select restoration project(s) that may be implemented to restore the trust resources and services injured by a given spill. The RRP Program's Final Programmatic Environmental Impact Statement (FPEIS) may be viewed in its entirety at http://www.losco.state.la.us/LOSCOuploads/RRPAR/la2395.pdf.

Trustees to seek public comment on the draft plan. The Trustees published the Phase II DERP/ER on November 6, 2012, and accepted comment on the draft through December 10, 2012. A public meeting was held on November 13, 2012 in Pensacola, Florida, to facilitate public review and comment on the plan.

The Trustees considered comments on the Phase II DERP/ER prior to finalizing the projects included in this Phase II ERP/ER. Summaries of comments received and Trustee responses are provided in Chapter 5 of this plan. Following publication of this Phase II ERP/ER, the Trustees will finalize agreements with BP regarding funding and offsets for the selected projects and proceed with implementation, subject to any remaining actions needed to comply with applicable state and federal laws.

## 1.10    Administrative Record

Pursuant to 15 C.F.R. § 990.45, the Trustees opened a publicly available Administrative Record (AR) for NRDA and restoration activities concurrently with the publication of the Notice of Intent to Conduct Restoration Planning. DOI is the lead federal Trustee for maintaining the AR, which can be found at http://www.doi.gov/deepwaterhorizon/adminrecord. Some of the state Trustees are also maintaining a state-specific AR (e.g., http://losco-dwh.com/AdminRecord.aspx). Information about project implementation will be provided to the public through the AR and other outreach efforts, including http://www.gulfspillrestoration.noaa.gov.

# CHAPTER 2    ENVIRONMENTAL SETTING – GULF OF MEXICO

## 2.1    Introduction

This chapter describes the general environment of the Gulf that provides the setting for the resources or services expected to benefit from the early restoration projects included in this Phase II ERP/ER. These are resources and services that, even at this early stage in the NRDA process, are known to be impacted as a result of the Spill. These impacts provide the nexus for the early restoration projects included in this Phase II ERP/ER. Gulf physical, ecological and socioeconomic resources are generally described in Chapter 2. Chapter 4 presents the NEPA review and other environmental compliance requirements.

## 2.2    Physical Environment

The Gulf ecosystem is made up of a complex, intricate array of interconnected natural resources. These natural resources provide a wide range of services to both the environment, itself, and to humans. The U.S. Gulf coastline extends across five states: Florida, Alabama, Mississippi, Louisiana and Texas. The overall watershed that drains into the Gulf extends over more than 50% of the continental United States (USGS and EPA, 2011 as cited in GCERTF, 2011). The Mississippi-Atchafalaya River Basin alone drains an estimated 40% of the continental United States (NOAA, 2011 as cited in GCERTF, 2011).

Coastal and marine environments of the Gulf include the intertidal zone, continental shelf, continental slope, and abyssal plain. The intertidal zone (also referred to as the foreshore or littoral zone) extends from mean lower low water to mean higher high water, and an upland area inward of mean higher high water. The upland area is not distinctly defined for this ERP/ER, but could include any area in the Gulf coast region potentially affected by a restoration project.

The continental shelf of the Gulf is seaward of the intertidal zone to the perimeter of the continental land mass. It can be divided into the inner and outer shelf environments. The extent of the continental shelf (miles from shoreline) and maximum depth at the shelf break varies throughout the basin. The inner continental shelf extends from mean lower low tide and is characterized by generally shallow waters and a gentle slope of a few feet per mile. The outer continental shelf is the deeper part of the shelf and extends to about a 650-foot depth contour.

Extending from the edge of the shelf to the abyssal plain, the outer continental slope is a steep area with diverse geomorphic features (canyons, troughs, and salt structures). The base of the slope in the Gulf occurs at a depth of about 9,000 feet. The Sigsbee Deep, located within the Sigsbee Abyssal Plain in the southwestern part of the basin, is the deepest region of the Gulf with a maximum depth ranging from about 12,000 to 14,000 feet (Figure 2).

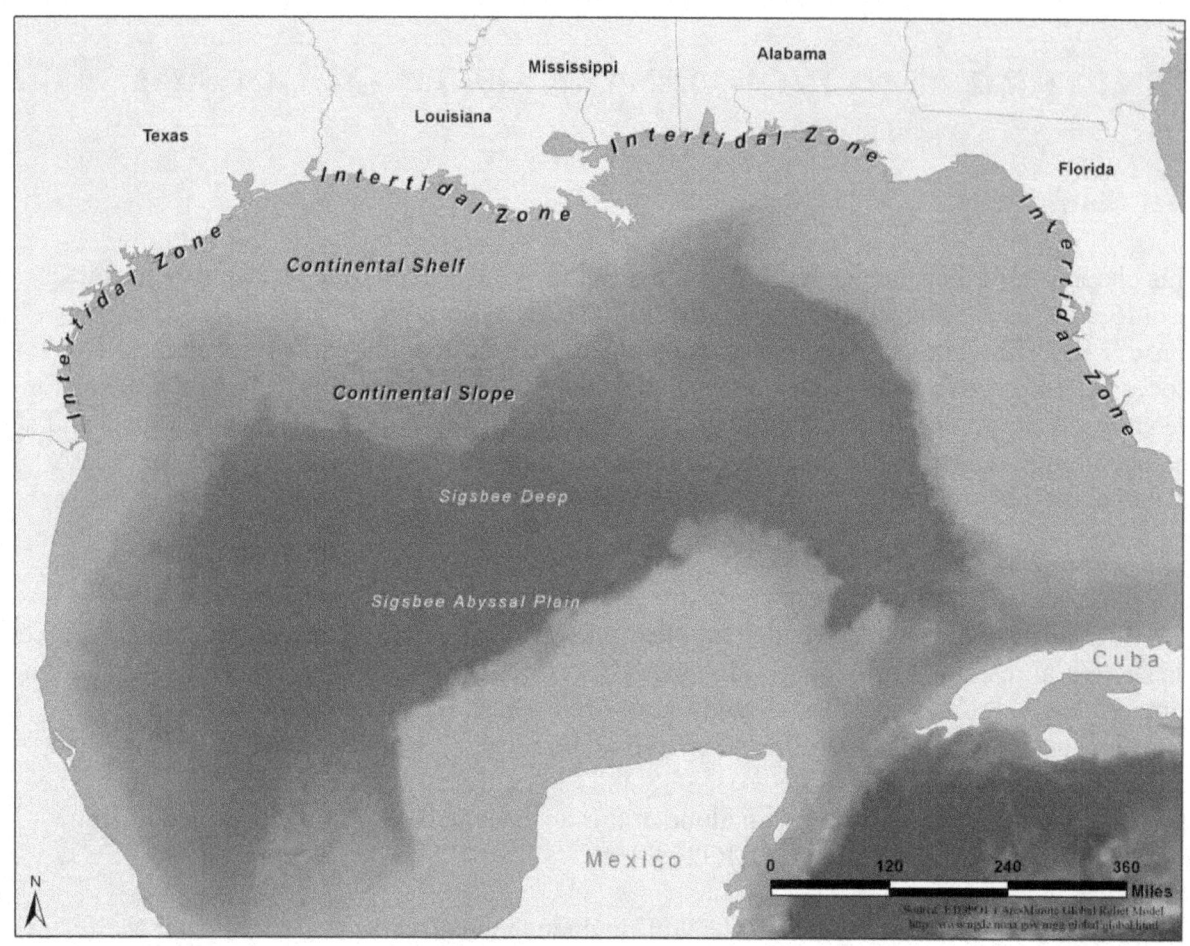

Figure 2. Gulf of Mexico.

## 2.3    Ecological Environment

The Gulf supports biologically diverse marine habitats and assemblages of species, including planktonic communities, bottom-dwelling organisms, deepwater corals, sponges, fish, birds, terrestrial and marine mammals, and other species and communities. The Gulf is also home to a number of coastal, marine, and freshwater fish and wildlife species listed as threatened or endangered, as well as several species of protected marine mammals.

The Gulf supports a variety of coastal and marine habitats, including wetlands, barrier islands, beaches, seagrass beds, and coral and oyster reefs. These interconnected habitats are essential for the diverse array of ecologically, commercially, and recreationally important species that occur in the Gulf. For example, intertidal wetlands and other nearshore habitats (which extend from Texas to Florida) provide foraging and nesting habitats for the numerous species of birds using the Mississippi Flyway, one of the most important migratory bird flyways in the world. These coastal areas also provide essential habitats for ecologically, commercially, and recreationally important species of fish and invertebrates.

Individually and collectively, these coastal and marine habitats are integral to the Gulf ecosystem, to both regional and national economies, and to the cultural fabric of the region and

16

the nation. Healthy Gulf Coast habitats and species provide a range of natural resource services including fisheries, food production, infrastructure protection, and recreational opportunities. Healthy Gulf Coast habitats also help to protect Gulf Coast communities, providing a line of defense against powerful storms, flooding and long term sea level rise.

### 2.3.1 Threatened, Endangered, and Candidate Species

Numerous species throughout the Gulf are listed as threatened or endangered through the Endangered Species Act of 1973 (ESA). These species are protected and as provided under ESA, federal consultations are required when environmental actions may affect these listed species or their designated critical habitat. Listed species potentially present in project areas are noted in Appendix A. Specific consideration of potential impacts to these species from these early restoration projects are further discussed in Chapter 4. ESA consultation correspondence will be available in the AR.

### 2.3.2 Essential Fish Habitat

The Magnuson-Stevens Act requires federal agencies to consult with NOAA Fisheries Service when any activity proposed to be permitted, funded, or undertaken by a federal agency may have adverse effects on designated essential fish habitat (EFH). EFH encompasses waterbodies, habitats, and substrates necessary for federally and regional fishery management council managed fish to complete various life history stages such as breeding, spawning, feeding or growth and survival to maturity. To comply with requirements of the Magnuson-Stevens Fishery Conservation and Management Act, the Trustees obtained and, where appropriate, are considering information on designated EFH in the Gulf from NOAA at http://www.habitat.noaa.gov/protection/efh/newInv/index.html, and from text descriptions in Fishery Management Plans also available at that site. Representative EFH categories are listed in Table 1.

Table 1. Representative categories of EFH identified in the Fishery Management Plan Amendment of the Gulf of Mexico Fishery Management Council.

| Estuarine Areas | Marine Areas |
| --- | --- |
| Estuarine emergent wetlands | Coral and coral reefs |
| Estuarine scrub/shrub mangroves | Non-vegetated bottoms |
| SAV | Artificial reefs |
| Oyster reef and shell banks | Water column |
| Intertidal flats | Live/Hard bottom |
| Palustrine emergent and forested wetlands | SAV |
| Mud/sand/shell/rock substrates | |
| Estuarine water column | |

## 2.4    Socioeconomic Environment

The Gulf is among the nation's most valuable and important ecosystems. The Gulf Coast and its natural resources are key components of the U.S. economy, producing 30% of the nation's gross domestic product in 2009 (NOAA, 2011 as cited in GCERTF, 2011). The region provides more than 90% of the nation's offshore oil and natural gas production (USEIA, n.d. as cited in GCERTF, 2011); 33% of the nation's seafood (Mabus, 2010 as cited in GCERTF, 2011); 13 of the top 20 ports by tonnage in the United States in 2009 (USACE, 2010 as cited in GCERTF, 2011); as well as regionally and nationally important tourism and recreational activities such as fishing, boating, beachcombing, and bird watching. These activities support more than 800,000 jobs (Mabus, 2010 as cited in GCERTF, 2011) across the region, providing a substantial economic input to Gulf communities and the nation. All of these industries depend on a healthy and resilient Gulf. The five U.S. Gulf Coast States, if considered an individual country, would rank seventh in global gross domestic product (NOAA, 2011 as cited in GCERTF, 2011).

## 2.5    Cultural Resources

The northern Gulf has a rich cultural heritage. Cultural resources are prehistoric, historic, or archaeological resources that have cultural significance and can include shipwrecks, historical buildings, monuments, and burial grounds. Cultural resources include historic properties listed in, or eligible for listing in the National Register of Historic Places (36 C.F.R. §60[a-d]). The National Historic Preservation Act of 1966 (NHPA), as amended (16 U.S.C. §470(f)), defines an historic property as "any prehistoric or historic district, site, building, structure, or object included in, or eligible for inclusion on the National Register [of Historic Places]." This includes significant properties of traditional religious and/or cultural importance to Indian tribes.

Historic properties include built resources (bridges, buildings, piers, etc.), archaeological sites, and Traditional Cultural Properties, which are significant for their association with practices or beliefs of a living community that are both fundamental to that community's history and a piece of the community's cultural identity. Although often associated with Native American traditions, such properties also may be important for their significance to ethnic groups or communities.

Historic properties also include submerged resources. Modern technology enables nautical archaeologists to recover data in areas previously inaccessible. The variety of shipping channels in the Gulf encompasses colonial and modern-day trade routes and activities. In addition, armed conflicts from colonial times to the 1940s have left indelible marks on the Gulf Coast. Shipwrecks can range from seventeenth century Spanish galleons to World War II-era German U-boats. Small pirogues or canoes may provide data on Native American or local history. Maritime archaeology includes but is not limited to the study of wrecks; wrecks encompass airplane and boat debris.

Bridges, shell middens, harbors, and villages can be submerged as a result of changing coastlines and other climatic activity. Approximately 19,000 years ago, global sea level was approximately 360 feet lower than present. During this time, large expanses of what is now the outer continental shelf were exposed as dry land. Twelve thousand years ago, the earliest date prehistoric human populations are known to have been in the Gulf Coast region (Aten, 1983, as cited in MMS, 2007), sea level would have been approximately 135 feet lower than present day levels (CEI,

1982, as cited in MMS, 2007). The location of the shoreline 12,000 years ago is roughly approximated by the 135-foot bathymetric contour. The continental shelf shoreward of this contour would have potential for prehistoric sites dating subsequent to 12,000 years ago. Since known prehistoric sites on land usually occur in association with certain types of geographic features, prehistoric sites should be found in association with those same types of features now submerged and buried on the continental shelf.

Geographic features that have a high potential for associated prehistoric sites include barrier islands and back barrier embayments, river channels and associated floodplains, terraces, levees and point bars, and salt dome features. A review of previously identified archaeological work in the vicinity of a project is critical to determining the scope of the archaeological identification effort. Areas subjected to previous extensive archaeological investigations may not warrant additional fieldwork. All previous work should be evaluated in consultation with State Historic Preservation Office and, if involved, a Tribal Historic Preservation Officer for reliability and accuracy.

## 2.6    Socioeconomic and Environmental Justice

To the greatest extent practicable, federal agencies must "identify and address, as appropriate, disproportionately high and adverse human health or environmental effects of its programs, policies, and activities on minority populations and low-income populations." Executive Order 12898 (Feb. 11, 1994). The Council on Environmental Quality (CEQ) issued guidance directing federal agencies to analyze the environmental effects, including human health, economic, and social effects, of their proposed actions on minority and low-income communities when required by NEPA. CEQ, Environmental Justice: Guidance under the NEPA, p. 25 (CEQ, 1997). CEQ defined members of minority populations to include: American Indian or Alaskan Native; Asian or Pacific Islander; Black, not of Hispanic origin; or Hispanic. Low income populations for this analysis were determined based on the U.S. Census Bureau 1999 poverty thresholds (USDOC, U.S. Census Bureau, 1999). Analyses in this Phase II ERP/ER comply with Executive Order 128898 and CEQ's guidance.

## 2.7    The *Deepwater Horizon* Oil Spill Natural Resource Damage Assessment

The Spill presents a complex threat to the interconnected organisms, habitats, and ecosystems of the Gulf. Unprecedented volumes of oil and dispersants were released into the environment and were transported in deepwater areas, the water column, along the ocean's surface, through coastal and nearshore areas, and onto shorelines. Figure 3 illustrates some of the various types of resources and services being evaluated as part of the *Deepwater Horizon* NRDA and provides a sense of the scope of investigations being done to fully evaluate the impacts of oil, dispersants, and other response actions on natural resources and the Gulf ecosystem.

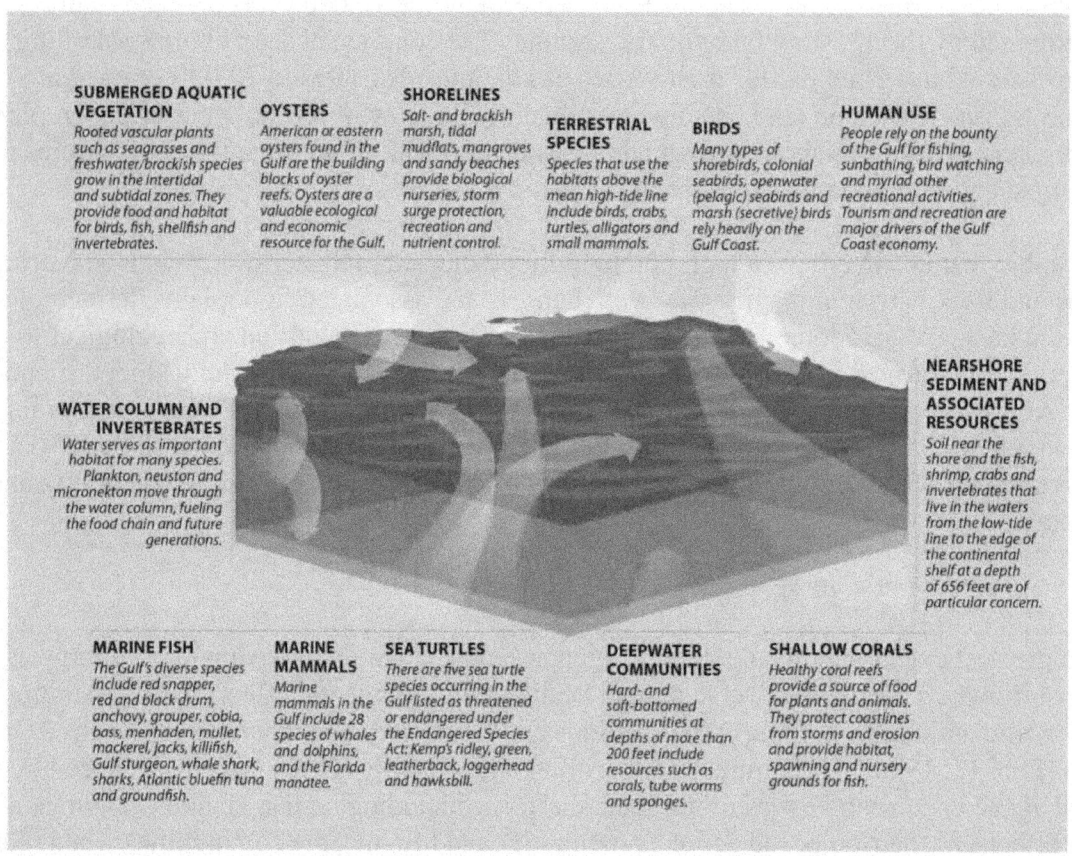

**SUBMERGED AQUATIC VEGETATION**
Rooted vascular plants such as seagrasses and freshwater/brackish species grow in the intertidal and subtidal zones. They provide food and habitat for birds, fish, shellfish and invertebrates.

**OYSTERS**
American or eastern oysters found in the Gulf are the building blocks of oyster reefs. Oysters are a valuable ecological and economic resource for the Gulf.

**SHORELINES**
Salt- and brackish marsh, tidal mudflats, mangroves and sandy beaches provide biological nurseries, storm surge protection, recreation and nutrient control.

**TERRESTRIAL SPECIES**
Species that use the habitats above the mean high-tide line include birds, crabs, turtles, alligators and small mammals.

**BIRDS**
Many types of shorebirds, colonial seabirds, openwater (pelagic) seabirds and marsh (secretive) birds rely heavily on the Gulf Coast.

**HUMAN USE**
People rely on the bounty of the Gulf for fishing, sunbathing, bird watching and myriad other recreational activities. Tourism and recreation are major drivers of the Gulf Coast economy.

**WATER COLUMN AND INVERTEBRATES**
Water serves as important habitat for many species. Plankton, neuston and micronekton move through the water column, fueling the food chain and future generations.

**NEARSHORE SEDIMENT AND ASSOCIATED RESOURCES**
Soil near the shore and the fish, shrimp, crabs and invertebrates that live in the waters from the low-tide line to the edge of the continental shelf at a depth of 656 feet are of particular concern.

**MARINE FISH**
The Gulf's diverse species include red snapper, red and black drum, anchovy, grouper, cobia, bass, menhaden, mullet, mackerel, jacks, killifish, Gulf sturgeon, whale shark, sharks, Atlantic bluefin tuna and groundfish.

**MARINE MAMMALS**
Marine mammals in the Gulf include 28 species of whales and dolphins, and the Florida manatee.

**SEA TURTLES**
There are five sea turtle species occurring in the Gulf listed as threatened or endangered under the Endangered Species Act: Kemp's ridley, green, leatherback, loggerhead and hawksbill.

**DEEPWATER COMMUNITIES**
Hard- and soft-bottomed communities at depths of more than 200 feet include resources such as corals, tube worms and sponges.

**SHALLOW CORALS**
Healthy coral reefs provide a source of food for plants and animals. They protect coastlines from storms and erosion and provide habitat, spawning and nursery grounds for fish.

Figure 3. Gulf of Mexico resources potentially affected by the *Deepwater Horizon* Spill.

The *Deepwater Horizon* NRDA includes assessment and evaluation of potential injuries to a wide array of natural resources, from the deep ocean to the coastlines of the northern Gulf. The injury assessment for the Spill is ongoing. Information continues to be collected to assess potential impacts to fish, shellfish, terrestrial and marine mammals, turtles, birds, and other sensitive resources as well as their habitats, including, but not limited to, wetlands, beaches, mudflats, bottom sediments, corals, and the water column. Lost human uses of these resources, such as recreational fishing, boating, hunting, and beachgoing, are also being assessed. Hundreds of scientists, economists, and restoration specialists have been and continue to be involved in these diverse NRDA activities.

Among the most readily observable impacts that have been a consequence of the Spill stem from the Gulf-wide response efforts aimed at reducing the short-term effects of oiling. These response efforts were undertaken at a massive scale, with nearly 50,000 responders active during the height of clean-up efforts. In addition, there were nearly 10,000 vessels involved in oil containment and removal, and millions of feet of absorbent and containment oil boom were deployed in an effort to reduce the amount of oil stranded along coastal shorelines. Although response efforts succeeded in reducing the amount of oil that was stranded on coastlines, these actions caused a number of unavoidable physical consequences on coastal resources, including smothering, trampling, removal, and disruptions in recreational use of beaches and waterways.

Natural resource impacts associated with response actions have not fully been quantified, and some may be ongoing.

Even at this early stage in the NRDA process, and even though the nature and extent of natural resource injuries and losses are still being assessed, some of the adverse effects of the Spill on natural resources or services have been observed and/or reasonably inferred, including due to response activities. Because this Phase II ERP/ER includes early restoration projects with a nexus to response injuries to beach habitat, the remainder of this chapter provides additional environmental information pertinent to this resource.

The Phase II ERP/ER includes two sandy beach habitat restoration projects discussed in Chapters 3 and 4 to restore injury to the habitat as a result of response activities.

The Gulf has hundreds of miles of sandy shoreline that are important both ecologically and economically. Sandy beaches are crucial habitat that support a variety of plant and animal species including federally or state listed sea turtles and beach nesting birds.

Response efforts were necessary and undertaken to prevent oil from coming ashore and to remove oil from beaches. These activities resulted in significant disturbance to nesting habitat on beaches. Response efforts physically impacted beaches as a result of effects from motorized vehicles, trampling, as well as removal of sand, vegetation, wrack, and shell, which are important biotic habitats. Continuous disturbance by response activities negatively affected habitat necessary for beach nesting birds as well as loggerhead sea turtles. Media coverage, aerial photography, Shoreline Cleanup and Assessment Teams (SCAT) records and other observational data include evidence of these physical impacts to beaches. Work to assess the full extent of these injuries is ongoing.

# CHAPTER 3 ALTERNATIVES, INCLUDING THE SELECTED ALTERNATIVE

Below we describe two alternatives that the Trustees considered for Phase II early restoration, the No Action alternative and the alternative selected by the Trustees.

## 3.1 Alternative A: No Action – Natural Recovery

Increased activity, including lights and equipment on the beach during the response, impacted the use of important nesting habitat by beach nesting birds and loggerhead sea turtles. Nesting habitat services were lost as a result of disturbance from lights and physical response activities in these nesting habitats. The projects propose to partially offset this injury by actively decreasing persistent and ongoing disturbance to beach habitat at specific sites. Under the No Action alternative, injury associated with disturbance of the nesting habitat resulting from the response will be left to natural recovery processes only.

Choosing this alternative, at this time, would not preclude analysis and implementation of different restoration activities at a later date. However, choosing No Action at this time would result in delaying protection and improvement of important nesting habitats injured by the Spill. The No Action alternative is used in this document as a basis for comparison of the effects from implementing the selected alternative. The baseline for comparison is defined as the current condition and expected future condition in the absence of the project(s). The Trustees have selected and will be proceeding with Alternative B described below to meet the goals articulated in Section 1.4, Purpose and Need for Early Restoration.

## 3.2 Alternative B: Selected Alternative – Phase II Early Restoration Projects

Based on analysis of the selection criteria set forth in OPA NRDA regulations, the Framework Agreement and additional Florida early restoration specific criteria, and consideration of public comment, the Trustees selected and intend to move forward with the two early restoration projects included in this alternative, namely: (1) Enhanced Management of Avian Breeding Habitat Injured by Response in the Florida Panhandle, Alabama, and Mississippi and (2) Improving Habitat Injured by Spill Response: Restoring the Night Sky. These projects are consistent with the goal of restoring or replacing ecological services lost due to the response to the Spill. The Trustees will finalize agreements for each project with BP (see Section 1.8), consistent with the Framework Agreement and the goal of implementing these projects prior to the 2013 nesting season to enhance important nesting habitats used by birds and sea turtles. Table 2 provides a brief overview of the selected projects.

### 3.2.1 Offsets Estimation Methodology for Projects

The Trustees used the Habitat Equivalency Analysis (HEA) method to estimate Offsets for these two early restoration projects. An overview of the Trustees' approach to estimating Offsets is outlined for each project.

Table 2. Early restoration projects selected under Alternative B.

| Project Title | Location (County and State) | Selected Restoration | Estimated Cost (including potential contingencies)[13] | Resources Benefitted |
|---|---|---|---|---|
| Enhanced Management of Avian Breeding Habitat Injured by Response in the Florida Panhandle, Alabama, and Mississippi | Florida: Escambia, Santa Rosa, Okaloosa, Walton, Bay, Gulf, and Franklin counties. Alabama: Bon Secour National Wildlife Refuge (NWR) in Baldwin and Mobile counties. Mississippi: Gulf Islands National Seashore (GUIS) – Mississippi District. | Symbolic fencing,[a] predator control, and stewardship around important nesting areas to prevent disturbance | $4,658,118 | Nesting and foraging habitat for beach nesting birds in Florida and on DOI lands in Alabama and Mississippi. |
| Improving Habitat Injured by Spill Response: Restoring the Night Sky | State-owned beaches within the boundaries of the Gulf State Park in Baldwin County, AL, and properties in Escambia, Santa Rosa, Okaloosa, Walton, Bay, Gulf, and Franklin counties, FL. | Reduce artificial lighting impacts on nesting habitat for loggerhead sea turtles | $4,321,165 | Nesting habitat for loggerhead sea turtles in Florida and state lands in Alabama. |

[a]See Figure 4 for an example of symbolic fencing.

Completed posted area (photo by Chris Burney)

Figure 4. Symbolic fencing protecting coastal habitat for beach nesting birds.

---

[13] Actual costs may differ depending on future contingencies, but will not exceed the amount shown without further agreement between the Trustees and BP.

HEA is commonly used in NRDAs to quantify changes in ecological services on a habitat basis (e.g., units of beach nesting habitat). When HEA is used to estimate restoration credits, anticipated ecological benefits resulting from the restoration action often are expressed in units that reflect the present (current) value of ecological benefits over a project's lifespan. For purposes of the early restoration projects included herein, the Trustees expressed HEA-estimated habitat benefits as "discounted service acre years" or DSAYs of the specific habitat types to be restored. For example, the Trustees estimated and expressed the present value of Offsets for the early restoration project to restore nesting habitat for beach nesting birds as DSAYs of nesting habitat for beach nesting birds. The Trustees considered a variety of project-specific factors when applying the HEA method to estimate the ecological benefits of restoration projects, including, but not limited to:

- The time at which ecological services from a restoration project begins to accrue;
- The rate of ecological service accrual over time;
- The time period over which ecological services will be provided;
- The quantity and quality of ecological services provided by the restored habitat or resource relative to those not affected by the Spill; and
- The size of the restoration action.

The methods used to estimate Offsets for these early restoration projects were implemented pursuant to the Framework Agreement. Offsets were negotiated with BP and reasonably reflect the estimated habitat service gains anticipated for each project. Neither the amount of the Offsets nor the methods of estimation are precedent for assessing the gains provided by any other projects either during the early restoration process or in the assessment of total injury. In the context of early restoration under the Framework Agreement, the Trustees are using best information and methodologies available in judging the adequacy of proposed restoration in satisfying OPA's mandates (see 15 C.F.R. Section § 990.25) while determining that agreements reached under the Framework Agreement are fair, reasonable, and in the public interest.

### 3.2.2    Overview of Selected Projects

Coastal sandy beach habitat was subject to disturbance from spill response activities. Gulf beaches provide critical ecosystem functions by providing nesting habitat to loggerhead sea turtles and beach nesting birds. Undisturbed stretches of coast are key components required for the life cycle of these species. The selected projects help address disturbance on beaches used for nesting by loggerhead sea turtles on Alabama state beaches and Florida beaches and beach nesting birds on federal beaches on Bon Secour NWR in Baldwin and Mobile Counties in Alabama; and on GUIS – Mississippi District in Mississippi.

#### 3.2.2.1 Enhanced Management of Avian Breeding Habitat Injured by Response in the Florida Panhandle, Alabama, and Mississippi

The Enhanced Management of Avian Breeding Habitat Injured by Response in the Florida Panhandle, Alabama, and Mississippi will reduce disturbance to beach nesting habitat for beach nesting birds in the project areas. The project involves three components. The first is placing symbolic fencing around sensitive beach nesting bird nesting sites to indicate the site as off-limits to people, pets, and other sources of disturbance (Figure 4). The second component is increased predator control to reduce disturbance and loss of eggs, chicks, and adult beach nesting birds at nesting sites. The final component is increasing surveillance and monitoring of posted nesting sites to minimize disturbance to beach nesting birds in posted areas.

### 3.2.2.1.1　Background and Project Description

When people and their pets enter nesting areas, beach nesting birds are disturbed, potentially resulting in nest abandonment, egg loss, and chick mortality. Posting important nesting areas effectively reduces human disturbance of nesting sites (Pruner et al., 2011). Enhanced Management of Avian Breeding Habitat Injured by Response in the Florida Panhandle, Alabama, and Mississippi will reduce disturbance to beach nesting habitat for beach nesting birds in important nesting areas on approximately 1,800-2,300 acres of state beaches in Escambia, Santa Rosa, Okaloosa, Walton, Bay, Gulf, and Franklin counties in Florida; federal beaches on Bon Secour NWR in Baldwin and Mobile Counties in Alabama; and on GUIS – Mississippi District in Mississippi (Figure 5; Table 3).

The project involves three components: (1) Placing symbolic fencing (signs and posts connected with rope) around sensitive nesting sites of beach nesting birds to indicate the site as off-limits to people, pets, and other sources of disturbance (e.g., Figure 4); (2) Increasing predator control to reduce disturbance and loss of eggs, chicks, and adult beach nesting birds at nesting sites, and (3) Increasing surveillance and monitoring of posted nesting sites to minimize disturbance to nesting habitat in posted areas. Fenced nesting habitat will be monitored to support adaptive management practices and responses (e.g., if beach nesting birds shift nesting site locations, posting materials will be relocated accordingly), and to gather data needed to quantitatively evaluate the effectiveness of the project. These actions would occur on approximately 1,800-2,300 acres of nesting habitat for beach nesting birds based on selected activities.

Predators (e.g., coyotes, raccoons, foxes, feral cats) of beach nesting birds, along with human activity, have degraded the overall quality of their nesting habitat. Therefore, predator control by non-lethal and lethal methods consistent with current management practices will be increased in Florida. Predator control will be implemented at the discretion of the land-managing agencies based on their evaluation of necessity and feasibility.
The project will be implemented in the following Florida counties: Escambia, Santa Rosa, Okaloosa, Walton, Bay, Gulf, and Franklin. In Alabama, the project will be implemented on Bon Secour NWR in Baldwin and Mobile Counties. In Mississippi, the project will be implemented on GUIS – Mississippi District. Figure 5 and Table 3 describe project locations.

Activities associated with this project will be ongoing for five years.

The total estimated project cost is $4,658,118.

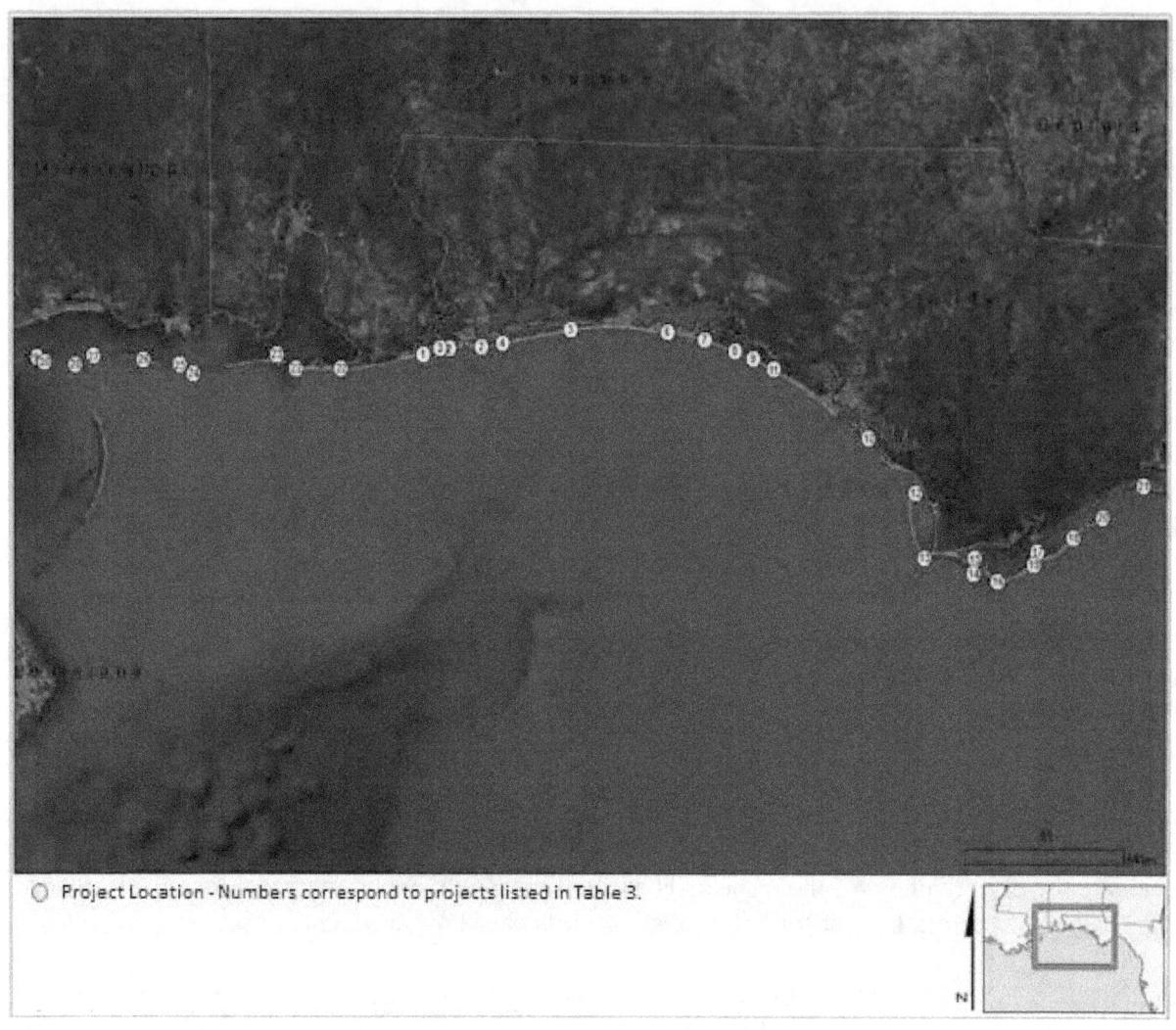

Figure 5. The Enhanced Management of
Avian Breeding Habitat Injured by Response project locations.

Table 3. Locations for Enhanced Management of Avian Breeding Habitat Injured by Response.

| Map Reference | Location County and State | Project Location Name |
|---|---|---|
| 1 | Escambia, FL | Central Perdido Key (Perdido Key State Park) |
| 2 | Escambia, FL | Eastern Perdido Key to western Santa Rosa Island (GUIS) |
| 3 | Escambia, FL | Big Lagoon State Park |
| 4 | Escambia, FL | Pensacola Beach |
| 5 | Santa Rosa, FL | Navarre Beach |
| 6 | Okaloosa, FL | Henderson State Park |
| 7 | Walton, FL | Top Sail Hill State Preserve |
| 8 | Walton, FL | Grayton Beach State Park |
| 9 | Walton, FL | Deer Lake State Park and Water Sound |
| 10 | Bay, FL | Shell Island to East Crooked Island |
| 11 | Bay, FL | Camp Helen State Park |
| 12 | Gulf, FL | St. Joseph Peninsula (St. Joseph Peninsula State Park) |
| 13 | Gulf, FL | Eglin Air Force Base – Cape San Blas |
| 14 | Franklin, FL | Flagg Island |
| 15 | Franklin, FL | St. Vincent NWR |
| 16 | Franklin, FL | Little St. George Island (Cape St. George State Reserve) |
| 17 | Franklin, FL | St. George Island Causeway (Apalachicola National Estuarine and Reserve) |
| 18 | Franklin, FL | St. George Island (St. George Island State Park) |
| 19 | Franklin, FL | St. George Island (portion outside of the State Park) |
| 20 | Franklin, FL | Dog Island |
| 21 | Franklin, FL | Alligator Point (Phipps Preserve) |
| 22 | Baldwin, AL | Ft. Morgan Peninsula, Bon Secour NWR |
| 23 | Mobile, AL | Little Dauphin Island, Bon Secour NWR |
| 24 | Jackson, MS | Petit Bois Island, GUIS |
| 25 | Jackson, MS | Spoil (Sand) Island, GUIS |
| 26 | Jackson, MS | Horn Island, GUIS |
| 27 | Harrison, MS | East Ship Island, GUIS |
| 28 | Harrison, MS | West Ship Island, GUIS |
| 29 | Harrison, MS | Cat Island, west end, GUIS |
| 30 | Harrison, MS | Cat Island, Smuggler's Cove, GUIS |

### 3.2.2.1.2    Selection Criteria

The goal of the Enhanced Management of Avian Breeding Habitat Injured by Response is to reduce disturbance to nesting habitat used by beach nesting birds. This nesting habitat improvement should improve successful nesting, hatching, and rearing of chicks (i.e., improve productivity). This important beach nesting habitat was negatively impacted during the Spill through the continued use of heavy equipment and presence of SCAT. Thus, the nexus to resources injured by the Spill is clear. See 15 C.F.R. § 990.54(a)(2). See also 6(a)-(c) of the Framework Agreement. Likelihood for success is very high based on success of similar efforts (Pruner et al., 2011). See 15 C.F.R. § 990.54(a)(3); and 6(e) of the Framework Agreement. The Enhanced Management of Avian Breeding Habitat Injured by Response can be conducted at a reasonable cost and can be implemented by the Trustees with minimal delay. See 15 C.F.R. § 990.54(a)(1); and 6(e) of the Framework Agreement. The project supports existing restoration initiatives and strategies and is consistent with anticipated long-term restoration needs and anticipated final restoration plans stemming from the Spill. See 6(d) of the Framework Agreement.

Protection of nesting habitat for beach nesting birds in Florida was suggested as a restoration measure during the public scoping meetings for the *Deepwater Horizon* Programmatic Environmental Impact Statement (PEIS) in Florida, submitted as a restoration project on the NOAA website (http://www.gulfspillrestoration.noaa.gov) and submitted to the State of Florida. In addition to meeting the evaluation criteria for the Framework Agreement and OPA, the selected project meets Florida's criteria that early restoration projects occur in the 8-county panhandle area where boom was deployed and that was impacted by the Spill or response to the Spill. These early restoration projects are consistent with recommendations made by Avissar et al. (2012) and Pruner et al. (2011).

### 3.2.2.1.3    Performance Criteria, Monitoring and Maintenance

Operation and maintenance activities will be required for this project. A supply of posting materials will need to be maintained. Symbolic fencing is subject to disturbance by storms and people and the need to re-post some areas is anticipated. Because of the dynamic nature of nesting site selection by beach nesting birds, regular observation of beach nesting birds will be needed to ensure that important areas are posted. Prior to, and after the project has been implemented, surveys of beach nesting bird habitat will be conducted in the project areas to record and evaluate data on changes in nesting/reproductive dynamics (e.g., levels of nesting effort and success).

The focal beach nesting bird species to be monitored for this project include the American oystercatcher, black skimmer, least tern, and snowy plover. These species have been opportunistically monitored by the FWC and various land management agencies (e.g., FDEP, DOD, NPS) for decades. However, in 1986, the FWC formed a non-game program and the new regional biologists began to more regularly conduct shorebird monitoring activities at sites that either lacked a strong land managerial presence, or where the land managers requested such assistance. These collaborative efforts have continued to the present day.

Monitoring of nests, eggs, chicks, and adult nesting shorebirds of these species will occur at all posted sites. The shorebird monitors shall follow the guidelines provided in the Florida Shorebird Alliance Guidelines for posting shorebird and seabird sites in Florida (Avissar et al., 2012). These guidelines include maps, directions to, and locations of all survey routes within project sites in Florida. Nesting data collected by the shorebird monitor in Florida will be entered into the Florida Shorebird Database at https://public.myfwc.com/crossdoi/shorebirds/index.html. All points of ingress and egress along survey routes will be determined and provided by the project manager. The shorebird monitor shall confine travel on the beach to these routes, and shall avoid walking or driving vehicles over dunes or dune vegetation. Moreover, the shorebird monitor shall comply with the "Best Management Practices for Operating Vehicles on the Beach" document found at http://flshorebirdalliance.org/pdf/FWC_beach-driving_BMPs.pdf.

Each shorebird nesting site will be monitored at weekly intervals beginning in mid-February (sites where snowy plovers nest) or beginning of May (sites where snowy plovers do not nest), and ending on all nesting sites by the end of August, or until all breeding activity has concluded (e.g., no active nests remain, and all juveniles and nesting birds have left the area), whichever is later. While monitoring, counts will be made of the location and number of shorebird nests, eggs, chicks, and nesting adults. Data will also be collected on the location, chronology, and number of eggs that hatch and the number of chicks that fledge per nest, and the number of nests, eggs, or chicks that are lost due to human (or pet) disturbances, storm events, or predators. Weekly counts of colonial nesting species (e.g., black skimmers and least terns) allow shorebird monitors to estimate peak numbers of nests, chicks, and flight-capable juveniles, which helps to better determine colony size, nesting success, and productivity. Similarly, weekly monitoring of nests of solitary nesting species (e.g., American oystercatchers and snowy plovers) also allows for better tracking of nest success and productivity of these species.

In addition, special attention will be given to the proximity of nests, eggs, chicks, and adult nesting shorebirds of these species to posted areas. If shorebirds are observed nesting, as evidenced by the presence of nests with eggs, chicks, or adults exhibiting nest defense behavior (e.g., "broken-wing" act) outside a posted area, or are no longer nesting within a posted area, the shorebird monitor will coordinate with the project manager within three (3) business days to discuss potential posting needs and (re)arrangements. If the shorebird monitor observes any unauthorized disturbance of nests, eggs, chicks, or nesting shorebirds (either within or outside posted areas) from people or their pets, the shorebird monitor may attempt to amiably resolve the situation.

### 3.2.2.1.4    Offset Methods Used and the Calculations Performed

For the purposes of negotiations of Offsets with BP in accordance with the Framework Agreement, the Trustees used HEA to estimate Offsets provided by the Enhanced Management of Avian Breeding Habitat Injured by Response. Offsets reflect units of DSAYs of nesting and foraging habitat for beach nesting birds, and would be applied against response injury to nesting and foraging habitat for beach nesting birds along the Florida coast and DOI lands in Alabama, and Mississippi.

In determining the DSAYs provided by the project, the Trustees considered a number of factors, including, but not limited to, the relative nesting habitat improvements provided by posting nesting sites and conducting predator control at various sites, the time period that posting and predator control would occur, and the anticipated acreage on which these activities would occur. Total estimated Offsets for the Enhanced Management of Avian Breeding Habitat Injured by Response are 1352 DSAYs of nesting and foraging habitat for beach nesting birds in Florida, applicable to response injuries to nesting and foraging habitat for beach nesting birds in Florida. Offsets are 54 DSAYs of nesting and foraging habitat for beach nesting birds on DOI lands in Alabama, applicable to response injuries to nesting and foraging habitat for beach nesting birds on DOI lands in Alabama. Offsets are 272 DSAYs of nesting and foraging habitat for beach nesting birds on DOI lands in Mississippi, applicable to response injuries to nesting and foraging habitat for beach nesting birds on DOI lands in Mississippi. These Offsets are reasonable for this resource and this project.[14]

### 3.2.2.2 Improving Habitat Injured by Spill Response: Restoring the Night Sky

The Improving Habitat Injured by Spill Response: Restoring the Night Sky project will reduce disturbance to nesting habitat for loggerhead sea turtles. The project involves multiple components: (1) Site-specific surveys of existing light sources for each targeted beach; (2) Coordination with site managers on development of plans to eliminate, retrofit, or replace existing light fixtures on the property or to otherwise decrease the amount of light reaching the loggerhead sea turtle nesting beach; (3) Retrofitting streetlights and parking lot lights; (4) Increased efforts by local governments to ensure compliance with local lighting ordinances; and (5) A public awareness campaign including educational materials and revision of the FWC Lighting Technical Manual (Witherington and Martin, 2000) to include Best Available Technology.

### 3.2.2.2.1 Background and Project Description

Loggerhead sea turtles (*Caretta caretta*) are listed as federally-threatened throughout their range, including the Northern Gulf of Mexico Recovery Unit (NGMRU), individuals of which nest on the Gulf coast from Franklin County in Florida west through Texas. A review of nest numbers through 2007 suggests the NGMRU of loggerheads is in a significant decline (NMFS and FWS, 2008; Witherington et al., 2009). Loggerhead sea turtles that nest on northeastern Gulf beaches are being evaluated as a distinct recovery unit of the larger Northwest Atlantic loggerhead distinct population segment (NMFS and FWS, 2008). Sandy beaches impacted by the Spill in this area provide important nesting habitat for this group of loggerheads.

---

[14] In the event that the Response Injury determination for nesting and foraging habitat for beach nesting birds in Florida, on DOI lands in Alabama and/or on DOI lands in Mississippi is characterized in the NRDA using a metric other than DSAYs of nesting and foraging habitat for beach nesting birds in Florida, on DOI lands in Alabama and/or on DOI lands in Mississippi, the Trustees agree to a NRDA Offset equal to sixty-six percent (66%) increase in the baseline productivity units for beach nesting birds in the project implementation area for the respective state. The productivity units shall be consistent with productivity units used in the NRDA beach nesting bird response injury quantification. Such Offsets shall be calculated by multiplying the baseline beach nesting bird productivity (as defined through the NRDA) in the project implementation area for the respective state by 1.66. If the offsets resulting from the projects exceed Response Injury to nesting and foraging habitat for beach nesting birds on DOI lands in Alabama or Mississippi, then any remaining credits (measured in the metric determined by the Trustees) are applicable to Response Injuries to nesting and foraging habitat for beach nesting birds in Alabama or Mississippi, respectively.

The Improving Habitat Injured by Spill Response: Restoring the Night Sky project will improve the quality of sandy beach as nesting habitat by addressing a pervasive negative impact, artificial lighting, to nesting females and hatchlings on the Gulf beaches. Artificial lights along beaches deter sea turtles from utilizing the area and modify essential behaviors, including migration to and from the beach and successful nesting. For example, a reduction in sea turtle nesting activity has been documented on beaches illuminated with artificial lights (Witherington, 1992; Witherington and Martin, 1996; Lohmann et al., 1997). In addition, artificial lights cause disorientation of individual animals (Salmon et al., 1992; Witherington, 1992).

The selected project will reduce disturbance to coastal nesting habitat for loggerhead sea turtles. The project will address beach habitat lighting issues at sites in Baldwin County, Alabama, and along conservation lands and nesting beaches in Escambia, Santa Rosa, Okaloosa, Walton, Bay, Gulf, and Franklin Counties in Florida (Figure 6).

Activities associated with this project will be ongoing for four years.

The estimated cost for this project is approximately $4,321,165.

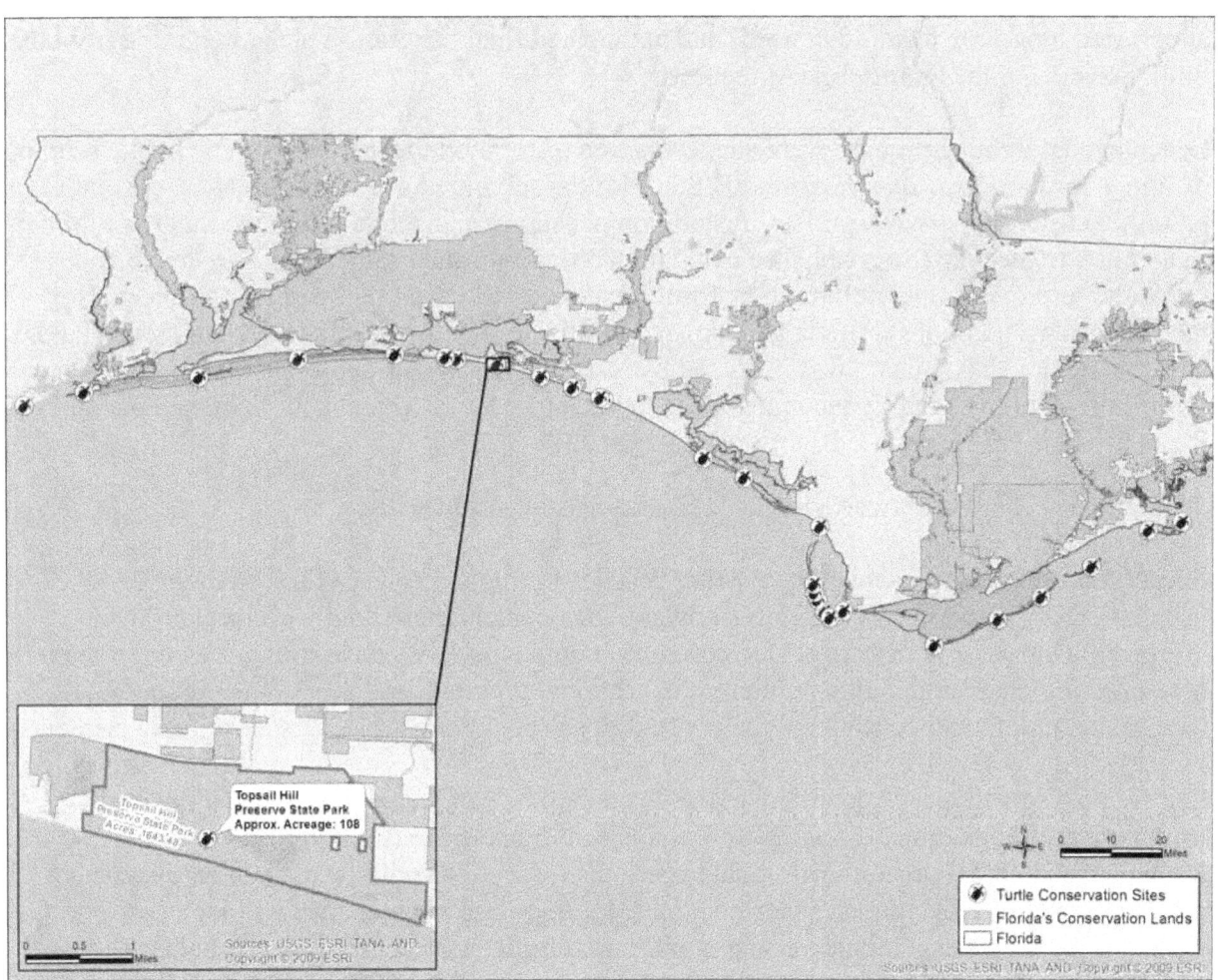

Figure 6. Improving habitat injured by spill response: Restoring the Night Sky project locations.

### 3.2.2.2.2 Selection Criteria

The goal of Improving Habitat Injured by Spill Response: Restoring the Night Sky is to offset the loss of ecological services due to response activities by improving the beach habitat for nesting and hatchling loggerhead sea turtles. During the Spill, heavy equipment was used and other response activities were conducted in the project areas around the clock. This 24-hour response necessitated the use of artificial lighting and dramatically increased human presence in beach habitat during nighttime hours. These activities caused disturbance and injury to the beach habitat and various types of impacts known to deter nesting loggerhead sea turtles (Witherington 1992). Thus, the nexus of the selected project to resources injured by the Spill is clear. (See 15 C.F.R. § 990.54(a)(2) and also 6(a)-(c) of the Framework Agreement.) Improving Habitat Injured by Spill Response: Restoring the Night Sky can be conducted at a reasonable cost and implemented by Trustees with minimal delay. See 15 C.F.R. § 990.54(a)(1) and (3); and 6(e) of the Framework Agreement. The project is technically feasible and utilizes proven techniques with established methods and documented results. Local, state, and federal agencies and non-governmental organizations have successfully implemented similar projects in Alabama and Florida. Therefore, the likelihood for success is very high based on success of similar efforts. The project supports existing restoration initiatives and strategies and is consistent with anticipated long-term restoration needs and anticipated final restoration plans stemming from the Spill. See 6(d) of the Framework Agreement.

Beach habitat lighting projects were suggested as a restoration measure during the public scoping meetings for the *Deepwater Horizon* PEIS in Florida, submitted as a restoration project on the NOAA website (http://www.gulfspillrestoration.noaa.gov) and submitted to the State of Florida. In addition to meeting the established evaluation criteria for the Framework Agreement and OPA, the Improving Habitat Injured by Spill Response: Restoring the Night Sky project also meets Florida's additional criteria that early restoration projects occur in the 8-county panhandle area that deployed boom and was impacted by response and SCAT activities for the Spill. This type of project is also highly recommended, and identified as a critical action, in the Federal Recovery Plan for Loggerhead Turtles (NMFS and FWS, 2008).

### 3.2.2.2.3 Performance Criteria Monitoring and Maintenance

Successful light management along nesting beaches for loggerhead sea turtles requires installation of appropriate lighting on landward development consistent with local ordinances, efforts by local governments to ensure continued compliance with local ordinances or protection measures, a focused and highly publicized educational program, and access to appropriate technical solutions and educational materials. The selected project includes all of these elements.

For each conservation site identified, assessments will be conducted of existing lights visible from the beaches on project areas as well as adjacent properties prior to lighting retrofits. Maintenance in the short-term will include periodic inspections with local code enforcement personnel to ensure lighting changes are retained. Long-term maintenance will include working with land managers to continue managing lighting retrofits as needed. After the lights are retrofitted, post-project assessments of the beach horizon will be conducted. Pre- and post-retrofit assessments will be compared to ensure that beach habitat lighting has been reduced.

Nine local governments in the Florida panhandle (six counties, three municipalities) have adopted lighting ordinances to facilitate protection of local sea turtle nesting beaches. Implementation and enforcement has been limited due to lack of funding, particularly when local resources were focused on Spill response efforts. As part of this project, local governments in the Florida panhandle will be provided with funds to increase staff time dedicated to inspections and compliance activities for the local lighting ordinances. To receive the additional funding, local governments will be expected to enter into a Memorandum of Agreement with appropriate Trustee(s) whereby commitments for education and enforcement of the local lighting ordinance will be specified. Compliance and enforcement tracking of the lights identified in the preliminary field inspections will be monitored to ensure corrective actions are being implemented. Local governments who agree to accept funds to improve compliance with local code enforcement efforts will be required to provide weekly or bimonthly summaries of all activities. Specific enforcement and compliance activities will be outlined in the official agreement for this activity; this agreement should also include specific targets for achieving compliance goals specified by the local government. Targets could include number of nighttime inspections, number of beachfront property owners contacted, number of notices provided to property owners (after initial contact does not achieve compliance with code requirements), number of violations pursued and resolved. Reporting of hours and travel shall be completed in accordance with all state purchasing and finance rules.

A public educational program will be developed and implemented in each of the seven western Florida panhandle coastal counties. Information on the importance of the loggerhead recovery unit in the Panhandle of Florida, on basic loggerhead sea turtle biology and nesting, and on lighting options to minimize impacts to the nesting beach will be provided via multiple media formats, including signage, public service announcements. The entity contracted to develop and implement the educational program will be required to develop survey techniques to test the effectiveness of the messaging and feedback from residents and visitors.

Monitoring of this multi-prong program will be implemented for each of the different components. A monthly summary of the number of lights removed or retrofit will be required for each public property. To document the reduction in the number of lights visible from the beach, annual surveys shall be required for each conservation land in addition to the pre- and post-project surveys. A requirement for such surveys will be included in the project agreements with state, local, and federal land managers and local governments. Project managers will use annual reports on lights to inform subsequent compliance and educational efforts.

Contractors involved in the public education campaign will be expected to provide routine updates on the status of the authorized educational programs, including number and format of educational activities as well as number of participants or other quantifiable metric. An important component of the public education campaign will include assessments of the efficacy of the specific activities on public knowledge and understanding of sea turtles and lights. Other monitoring activities may include surveys mailed to properties surrounding the parks or surveys conducted during beach festivals and other events in the target counties. Education programs will be required to utilize social media tools including Twitter and Facebook and to provide information on the number of "hits" or participants in a weekly summary.

This project will include on-beach assessments of habitat quality conducted prior to, during, and at the conclusion of the project. Empirical or categorical assessments of the overall "darkness" of the beach, the presence of natural landward silhouettes, the slope of the beach, and amount of disturbance will be considered.

### 3.2.2.2.4    Offset Methods Used and the Calculations Performed

For the purposes of negotiations of Offsets with BP in accordance with the Framework Agreement, the Trustees used HEA to estimate Offsets provided by Improving Habitat Injured by Spill Response: Restoring the Night Sky. Offsets reflect units of DSAYs of nesting habitat for nesting loggerhead sea turtles, and will be applied against response injury to nesting habitat for loggerhead sea turtles along the Florida and Alabama coast injured by the Spill response as determined by the Trustees' injury assessment. In determining DSAYs for this project, the Trustees considered a number of factors, including, but not limited to, the relative habitat benefits provided by reducing artificial lighting on loggerhead nesting beaches, the anticipated performance of the lights over time, and the potential number of acres of loggerhead nesting habitat that would be improved by the project. Total estimated Offsets for Improving Habitat Injured by Spill Response: Restoring the Night Sky are 1053 DSAYs of sea turtle nesting habitat in Florida, applicable to response injuries to sea turtle nesting habitat in Florida. Offsets are 31 DSAYs of sea turtle nesting habitat in Alabama, applicable to response injuries to sea turtle nesting habitat in Alabama. These Offsets are reasonable for this resource and this project.[15]

---

[15] In the event that the Response Injury determination for sea turtle nesting habitat in Florida and/or Alabama is quantified in the NRDA using a metric other than DSAYs of sea turtle nesting habitat in Florida and/or Alabama, the Trustees agree to a NRDA Offset applicable to Florida or Alabama, respectively, equal to a ten percent (10%) increase in the number of hatchlings reaching the sea over 10 years from a historic baseline that accounts for the sea turtle nesting activity in the project implementation areas in the respective states. Such Offsets shall be calculated by multiplying the baseline nest emergence success or the baseline rate of hatchlings reaching the water (as defined through the NRDA) by 1.10. If the offsets resulting from the projects exceed Response Injury to sea turtle nesting habitat in Florida or Alabama, respectively, and if the Trustees measure any such remaining credits using a metric other than DSAYs, the Trustees agree to a Natural Resource Damage Offset applicable to Florida or Alabama, respectively, for any such remaining credits using the foregoing 10% rate of increase and 10 year timeframe.

# CHAPTER 4   ENVIRONMENTAL COMPLIANCE

The Trustees selected the two early restoration projects described in Chapter 3 of this Phase II ERP/ER. These projects address coastal habitat and its services injured by the Spill response. The "Enhanced Management of Avian Breeding Habitat Injured by Response in the Florida Panhandle, Alabama, and Mississippi" project is located in Florida, Alabama, and Mississippi. The "Improving Habitat Injured by Spill Response: Restoring the Night Sky" project is located in Florida and Alabama.

This chapter addresses the Trustees' compliance with NEPA, 42 U.S.C. §§ 4321 *et seq.*, and other environmental planning requirements for these projects. The Trustees combined these two projects into one early restoration plan under OPA, however, for purposes of NEPA, the Trustees considered each project separately because they have independent utility.[16]

Under NEPA, federal agencies must consider and disclose the environmental impacts of major federal actions, such as undertakings on federal lands, issuing permits, or providing funding. Federal agencies may categorically exclude certain actions from further NEPA analysis because such actions characteristically do not have a significant effect on the human environment, individually or cumulatively. An EA is prepared for actions that do not qualify for a CX, and is a concise public document that provides information to determine if an action involves significant environmental impacts. Where a specific action or set of actions has already been the subject of an EA analysis by another federal agency, a federal Trustee may adopt and rely on that prior EA in making its own NEPA determinations for the proposed action. If an EA does not lead to a FONSI and instead identifies a potential for significant environmental impacts, then the agency must prepare an EIS.

Each of these projects is justified and would be undertaken regardless of whether the other project would be undertaken, and regardless of whether any additional future restoration is undertaken. The Trustees developed, evaluated, and negotiated with BP each of the projects independent from the others. While the Trustees intend to complete one billion dollars in early restoration projects under the Framework Agreement, additional restoration projects are subject to future negotiations. Therefore, each project, including their direct, indirect, and cumulative impacts, has been analyzed separately under NEPA.

As discussed below, the Improving Habitat Injured by Spill Response: Restoring the Night Sky project falls within a FWS CX and no further NEPA analysis is required. The predator control portion of the Enhanced Management of Avian Breeding Habitat Injured by Response in the

---

[16] NEPA provides that actions that are connected or dependent on other actions to be analyzed together in one NEPA analysis. Actions are considered connected if: (1) they automatically trigger other actions which may require an EIS(s), (2) they cannot or will not proceed unless other actions are taken previously or simultaneously, or (3) they are interdependent parts of a larger action and depend on the larger action for their justification. The projects do not fit the description of connected actions in 40 C.F.R. § 1508.25. First, to the best of the Trustees' knowledge, none of the projects would automatically trigger other actions which may require an EIS(s). Second, each of the projects represents a whole project and their performance does not depend on the previous or simultaneous performance of any other action. Third, the projects are not an interdependent part of a larger action.

Florida Panhandle, Alabama, and Mississippi project is the subject of a prior Final EA analysis by another federal agency (USDA), which FWS and NPS have reviewed and are adopting for activities within the predator control portion of this project. Additional Enhanced Management of Avian Breeding Habitat Injured by Response in the Florida Panhandle, Alabama, and Mississippi project activities that are not included in that prior EA (i.e., placing symbolic fencing and increasing surveillance, training, outreach, and monitoring) – all activities which would normally be categorically excluded – are analyzed below as part of the adoption process. Therefore, for purposes of this project, this Phase II ERP/ER supplements the adopted EA. Below is an overview of CXs and the EA adoption process, followed by a discussion of the CXs and the adopted EA as applicable to each of the projects.

## 4.1    Overview of Categorical Exclusions

NEPA requires Federal agencies to analyze their proposed actions to determine if they could have significant environmental effects. Over time, through study and experience, agencies may identify activities that do not need to undergo detailed environmental analysis in an EA or an EIS because the activities do not individually or cumulatively have a significant effect on the human environment. Agencies can define categories of such activities, called CXs, in their NEPA implementing procedures, as a way to reduce unnecessary paperwork and delay. The DOI NEPA Regulations contain Departmental CXs (43 C.F.R. §46.210) and individual DOI bureaus maintain additional CXs [516 DM 8.5 (FWS), 516 DM 12.5 (NPS)].

If a DOI bureau determines that a proposed activity fits within the description of one or more CXs, no additional NEPA review is required and the bureau can proceed with the activity without preparing an EA or EIS. CXs are an essential tool in facilitating NEPA implementation and concentrating environmental reviews on instances of potential impacts. A CX is a form of NEPA compliance, without the need for further project-by-project analysis through an EA. CXs are not exemptions or waivers of NEPA review; they simply give rise to a different type of NEPA review.

The DOI NEPA Regulations require that before a CX is used a list of "extraordinary circumstances" be reviewed for applicability (43 C.F.R. § 46.215). Extraordinary circumstances are factors or circumstances in which a normally excluded action may have a significant environmental effect that then requires further analysis in an EA or EIS. When no extraordinary circumstances exist a CX may be applied and the NEPA process ends without need for further review.

## 4.2    Overview of Adoption of Another Agency's Environmental Assessment

Federal agencies are encouraged to coordinate and take appropriate advantage of existing NEPA documents and studies, including adoption and incorporation by reference. Under CEQ NEPA Regulations (40 C.F.R. § 1506.3), DOI NEPA Regulations (43 C.F.R. § 46.120), and individual DOI bureau NEPA procedures, a DOI bureau can adopt another federal agency's EA to streamline the NEPA compliance process.

Pursuant to these authorities and FWS and NPS NEPA procedures, prior to adopting another federal agency's EA, the decision maker must independently evaluate the EA to ensure that the adopted document adequately reflects significant issues raised during scoping, adequately addresses public comments on the EA, includes actions and alternatives to be considered by the decision maker, and adequately addresses the impacts of the proposed action and alternatives. Public involvement requirements must also be met before FWS and NPS can adopt another agency's EA. The decision maker must prepare his/her own FONSI which acknowledges the origin of the EA and takes full responsibility for the scope and content.

## 4.3 Enhanced Management of Avian Breeding Habitat Injured by Response in the Florida Panhandle, Alabama, and Mississippi

### 4.3.1 Purpose and Need for Restoration

Bird nesting and breeding habitat was exposed to oil and dispersants and/or affected by response activities undertaken to prevent, minimize, or remediate oiling from the Spill. Under OPA, the Trustees act on behalf of the public to restore, rehabilitate, replace, or acquire the equivalent of natural resources injured and associated service losses as a result of the Spill. The beaches of the Florida Panhandle, Alabama, and Mississippi barrier islands provide vital nesting-season habitat for beach nesting birds. During spill surveillance and clean-up efforts, adult birds and their nests were repeatedly disturbed during the nesting season, particularly species with special status under various state authorities, including the snowy plover, Wilson's Plover, least tern, American oystercatcher, black skimmer, and brown pelican.

The purpose of this project under OPA and the Framework Agreement is to address response injuries to nesting habitat incurred during the Spill. This project would improve the quality and functioning of nesting habitat used by Gulf beach nesting birds in the project area.

### 4.3.2 Project Scope

The project would be implemented in the following Florida counties: Escambia, Santa Rosa, Okaloosa, Walton, Bay, Gulf, and Franklin (see Table 3 for a consolidated site summary). In Alabama, the project would be implemented on Bon Secour NWR in Baldwin and Mobile Counties. In Mississippi, the project would be implemented on GUIS – Mississippi District. Similar work in Florida has demonstrated that the management activities included in this project can be successful in improving critical nesting habitat.

Project partners are the FDEP, DOI, DOD, local governments, and NOAA.

The project would enhance affected habitats for beach nesting birds by implementing a coordinated management program over the next five years. This project would address the most significant needs associated with these habitats within the project locations. Management actions to improve these habitats would include the following:

1) Placing symbolic fencing around sensitive bird nesting sites to indicate the site as off-limits to people, pets, and other sources of disturbance;
2) Increasing surveillance and efficacy of posted nesting sites with increased training, outreach, and monitoring by the FWC, FDEP, NPS and FWS biologists and staff to minimize disturbance to nesting birds in posted areas;
3) Increased predator control to reduce disturbance of eggs, chicks, and adult birds at nesting sites in Florida.

Posted nesting sites would be monitored to support adaptive management practices/responses (e.g., if birds shift nesting site locations, posting materials would be relocated accordingly), and to gather the data needed to quantitatively evaluate the effectiveness of the management actions.

These actions would occur on approximately 1,800-2,300 acres of nesting habitat for beach nesting birds (range based on management activities being proposed).

Prior to, and after project implementation, surveys of nesting sites for beach nesting birds would be conducted in the project areas to record and evaluate data on changes in nesting/reproductive dynamics (e.g., levels of nesting effort and success).

The project supports existing restoration initiatives and strategies and is consistent with anticipated long-term restoration needs and anticipated final restoration plans stemming from the Spill. See 6(d) of the Framework Agreement.

### 4.3.3   Predator Control Activities

Recent increases in predators (e.g., coyotes, raccoons, foxes, feral cats) of beach-nesting birds, along with human activities, have degraded the overall quality of their nesting habitat. Therefore, one aspect of this project is to increase predator control in Florida through additional contracting with USDA, Animal and Plant Health Inspection Service (APHIS), Wildlife Services (WS). This aspect of the project has been evaluated by FWS and NPS under NEPA through a Final EA prepared by USDA/WS, for which FWS was a cooperating agency.

Predator control has been implemented by WS for many years in Florida as a successful method of improving the quality of beach nesting habitats for birds. As the prior federal proponent of this type of activity, WS completed an EA and issued a FONSI for implementation of these activities under Cooperative Agreements.

For this project, WS would conduct the same predator control activities in accordance with Cooperative Agreements as described within the existing EA and at the discretion of the land-managing agencies based on their evaluation of necessity and feasibility. The environmental impacts of the predator control component of this early restoration project are analyzed wholly within this prior EA, and it is reviewed and updated as appropriate. The WS EA and FONSI are included in Appendix D and are incorporated herein.

FWS and NPS have independently evaluated the WS EA and each believes that it satisfies all of the requirements for adoption. Because the potential impacts of the predator control activities in the project are sufficiently analyzed in the WS EA and DOI/FWS was a cooperating agency in its preparation, FWS and NPS are adopting the EA and relying on that EA in making NEPA determinations for the project.

### 4.3.4   Other Project Activities

The WS EA does not address the potential environmental effects of some of the activities that are part of the Enhanced Management of Avian Breeding Habitat Injured by Response in the Florida Panhandle, Alabama, and Mississippi project, i.e., placing symbolic fencing around sensitive bird nesting sites and increasing surveillance and efficacy of posted nesting sites with increased training, outreach, and monitoring. Due to the aggregation of these other project activities with predator control activities in the project, FWS and NPS considered whether all of the activities included in the Enhanced Management of Avian Breeding Habitat Injured by Response in the Florida Panhandle, Alabama, and Mississippi project would have a significant effect on the quality of the human environment.

If the placement of symbolic fencing and/or increased training, outreach, and monitoring activities had been proposed alone and not in combination with predator control activities in Florida, they would have been categorically excluded under one or more of the following FWS and NPS CXs (listed at 516 DM 8.5 and 516 DM 12.5, respectively):

- **516 DM 8.5A(2)** Personnel training, environmental interpretation, public safety efforts, and other educational activities, which do not involve new construction or major additions to existing facilities.
- **516 DM B(3)** The construction of new, or the addition of, small structures or improvements, including structures and improvements for restoration of wetland, riparian, in-stream, or native habitats, which result in no or only minor changes in the use of the affected local area. The following are examples of activities that may be included.
    i. The installation of fences.
    ii. The construction of small water control structures.
    iii. The planting of seeds or seedlings and other minor revegetation actions.
    iv. The construction of small berms or dikes.
    v. The development of limited access for routine maintenance and management purposes.
- **516 DM 8.5B(11)** NRDA restoration plans, prepared under sections 107, 111, and 122(j) of the Comprehensive Environmental Response Compensation and Liability Act (CERCLA); section 311(f)(4) of the Clean Water Act; and the OPA; when only minor or negligible change in the use of the affected areas is planned.
- **516 DM 12.5C(20)** Construction of fencing enclosures or boundary fencing posing no effect on wildlife migrations.

As previously discussed, actions that are subject to an agency's CX have previously been determined by that agency through study and experience to have no significant effect on the human environment, individually or cumulatively. However, due to a potential for causing significant effects when combining a series of actions that individually do not cause significant effects, FWS and NPS did not rely on these CXs; rather they supplemented the WS EA with additional analysis of the potential impacts of the project.

In supplementing the WS EA, both FWS and NPS analyzed the potential impacts of the entire project, as demonstrated in the appended environmental analysis documentation (see Appendix E). FWS and NPS evaluated whether implementing the project might result in significant effects on any of a range of physical and natural resources (e.g., air quality; water quality; wetlands; threatened and endangered species; other wildlife or wildlife habitat; visitor experience; socioeconomics; etc.). Placing symbolic fencing around sensitive bird nesting sites and increasing training, outreach, and monitoring, in addition to predator control activities, would not result in a significant impact, individually or cumulatively, on the quality of the human environment.

On balance, the project has positive effects that are consistent with long-term planning goals and contribute beneficially to avian nesting habitat in Florida, Alabama, and Mississippi. Additionally, all effects are local to the project areas, geographically disparate, and are not expected to overlap the activities or locations of other projects that the Trustees have approved as early restoration, including the eight projects contained in the prior Phase I ERP/EA. Together, the adopted WS EA, incorporated by reference and appended, and the appended environmental analysis documentation, is a Supplemental EA that satisfies the NEPA compliance requirement for this project.

### 4.3.5 Compliance with Other Laws

A consultation for this project under Section 7 of the ESA will be completed prior to project implementation. This project would be implemented in accordance with all applicable laws and regulations concerning the protection of threatened and endangered species and their habitats.

A complete review of this project under Section 106 of the NHPA will be completed prior to project implementation. This project would be implemented in accordance with all applicable laws and regulations concerning the protection of cultural and historic resources.

Under the Coastal Zone Management Act of 1972, the selected projects must be consistent to the maximum extent practicable with the federally-approved coastal management programs for the states in which the projects are to be conducted. The Federal Trustees submitted a consistency determination for this project for appropriate state reviews coincident with public review of the Phase II DERP/ER. Each state responded and concurred with the federal determination for the project for purposes of finalizing this early restoration plan.

The Magnuson-Stevens Act requires federal agencies to consult with NOAA Fisheries Service when any activity proposed to be permitted, funded, or undertaken by a federal agency may have adverse effects on designated EFH. In estuaries, EFH includes intertidal flats that are also used as foraging areas by shorebirds during low tides. However, no project activities beyond monitoring will be implemented in intertidal flats. The Trustees have therefore determined that the project does not have the potential to impact any designated EFH.

### 4.3.6 Summary

Because the Enhanced Management of Avian Breeding Habitat Injured by Response in the Florida Panhandle, Alabama, and Mississippi project only involves the seasonal placement of symbolic fencing around sensitive bird nesting sites; increasing surveillance and efficacy of posted nesting sites with increased training, outreach, and monitoring by FWC, FDEP, NPS, and FWS biologists and staff; and increased predator control which has been adequately analyzed in an existing EA, and would result in only minor or negligible change in the use of the project areas, FWS and NPS have adopted the existing WS EA for this project and this Phase II ERP/ER as a supplement to the WS EA for this project.

## 4.4 Improving Habitat Injured by Spill Response: Restoring the Night Sky

### 4.4.1 General Project Information

This project would improve the quality of sandy beach as habitat for loggerhead sea turtles. Artificial lights along beaches deter sea turtles from utilizing the area and modifying essential behaviors, including migrating, sheltering, nesting, and foraging. For example, a significant reduction in sea turtle nesting activity has been documented on beaches that are illuminated with artificial lights. In addition, artificial lights cause disorientation of individual animals.

Retrofitting existing street lights to reduce visibility from the beach and efficiently focus the illumination where it is most needed is a common regional practice to enhance the value of beach habitat. This project will seek such an enhancement to be achieved by reducing the amount of light cast onto beaches from anthropogenic sources within and adjacent to state, federal, and local lands in the Florida Panhandle and Gulf State Park property in Baldwin County, Alabama.

### 4.4.2 Project Scope

This project would be implemented in Baldwin County, AL, and Escambia, Santa Rosa, Okaloosa, Walton, Bay, Gulf, and Franklin Counties, FL. Project partners are the FDEP, Florida and Alabama local governments, Eglin Air Force Base, Tyndall Air Force Base, Baldwin EMC, and Gulf State Park.

As integral parts of this Improving Habitat Injured by Spill Response: Restoring the Night Sky project, the Trustees would conduct site-specific surveys of existing light sources for each targeted beach; coordinate with site managers on development of plans to eliminate, retrofit, or replace existing light fixtures on the property or to otherwise decrease the amount of light reaching the sea turtle nesting beach; conduct a before-and-after lighting impact assessment; and revise the FWC Lighting Technical Manual (Witherington and Martin, 2000) to include Best Available Technology. Similar, successful lighting retrofit efforts have been conducted for decades.

The project supports existing restoration initiatives and strategies and is consistent with anticipated long-term restoration needs and anticipated final restoration plans stemming from the Spill. See 6(d) of the Framework Agreement.

### 4.4.3 Categorical Exclusions

After undergoing NEPA review, the Trustees determined that the project meets FWS CXs. A NEPA Compliance Checklist (FWS Form 3-2185) documents the CXs and demonstrates that none of the "extraordinary circumstances" that require exceptions to CXs (43 C.F.R. § 46.215) apply to these activities (Appendix F).

The applicable CXs from 516 DM 8.5 (FWS) are listed below:

- **516 DM 8.5A(2)** Personnel training, environmental interpretation, public safety efforts, and other educational activities, which do not involve new construction or major additions to existing facilities.
- **516 DM 8.5B(2)** The operation, maintenance, and management of existing facilities and routine recurring management activities and improvements, including renovations or replacements which result in no or only minor changes in the use, and have no or negligible environmental effects on-site or in the vicinity of the site.
- **516 DM 8.5B(11)** NRDA restoration plans, prepared under sections 107, 111, and 122(j) of CERCLA; section 311(f)(4) of the Clean Water Act; and the OPA; when only minor or negligible change in the use of the affected areas is planned.

Due to the applicability of these CXs, no additional NEPA analysis for this project is required at this time.

### 4.4.4 Compliance with Other Laws

A complete consultation for this project under Section 7 of the ESA will be completed prior to project implementation. This project would be implemented in accordance with all applicable laws and regulations concerning the protection of threatened and endangered species and their habitats.

A complete review of this project under Section 106 of the NHPA will be completed prior to project implementation. This project would be implemented in accordance with all applicable laws and regulations concerning the protection of cultural and historic resources.

As selected, this project does not include the replacement of fixtures, if any, that are listed or eligible for listing on the National Register of Historic Places. If the Section 106 review process yields information that necessitates modifying the project, the project will be re-evaluated as appropriate in accordance with all applicable laws.

Pursuant to the Coastal Zone Management Act of 1972, the projects must be consistent to the maximum extent practicable with the federally-approved coastal management programs for the states in which the projects are to be conducted. The Federal Trustees submitted a consistency determination for this project for appropriate state reviews coincident with public review of the Phase II DERP/ER. Each such state responded and concurred with the federal determination for the project for purposes of finalizing this early restoration plan.

The Magnuson-Stevens Act requires federal agencies to consult with NOAA Fisheries Service when any activity proposed to be permitted, funded, or undertaken by a federal agency may have adverse effects on designated EFH. In estuaries, EFH includes intertidal flats that are also used as foraging areas by shorebirds during low tides. However, no project activities beyond monitoring will be implemented in intertidal flats. The Trustees have therefore determined that the project does not have the potential to impact any designated EFH.

### 4.4.5 Summary

Because the Improving Habitat Injured by Spill Response: Restoring the Night Sky project only involves retrofitting of streetlights and parking lot lights; site-specific surveys of existing light sources for each targeted beach; coordination and development of plans; and lighting impact assessments and technical manual revisions, and would result in only minor or negligible change in the use of the project areas, FWS has determined to apply CXs to this project.

### 4.5 Conclusion

Overall, the selected projects would enhance habitats that are important for nesting of beach nesting birds and for loggerhead sea turtles in the project areas. The Trustees have determined that the Improving Habitat Injured by Spill Response: Restoring the Night Sky project qualifies for CXs and that there are no extraordinary circumstances that might cause significant environmental effects. Therefore, no further NEPA analysis of this project is necessary. With respect to the Enhanced Management of Avian Breeding Habitat Injured by Response in the Florida Panhandle, Alabama, and Mississippi project, the potential impacts of predator control activities are analyzed in an existing EA, for which FWS was a cooperating agency. That EA is adopted by FWS and NPS. The remaining project activities, though normally categorically excluded, would have no potential for significant effect on the quality of the human environment when considered in conjunction with the predator control activities. Therefore, no need for an EIS has been identified.

Since project scope, environmental conditions, and regulatory requirements can change over time, any use of CXs will be reviewed for continued applicability prior to and during project implementation.

# CHAPTER 5   PUBLIC COMMENT ON DRAFT PHASE II EARLY RESTORATION PLAN AND ENVIRONMENTAL REVIEW AND RESPONSES

The public comment period for the Phase II DERP/ER opened November 6, 2012 and closed December 10, 2012. During this time, the Trustees hosted a public meeting in Pensacola, Florida, at which the Trustees accepted written comments and verbal comments. In addition to comments provided at the public meeting, the Trustees received web-based submissions, emailed submissions, and mailed-in submissions.

Following the comment period, the Trustees reviewed all submissions. Similar or related comments were then grouped and summarized for purposes of response. All comments submitted during the period for public comment were reviewed and considered by the Trustees prior to finalizing the Phase II ERP/ER. All comments submitted are represented in the summary comment descriptions listed in this chapter.

Comments received were both general in nature as well as directed toward specific aspects of the two projects. All public comments will be included in the AR.

## 5.1    General Comments

Comments that were not specific to particular projects, but generally applicable to the public comment process, project selection, residual contamination, project implementation, monitoring, new project ideas and other 'general' topics are addressed in Section 5.1. Comments specific to particular projects are addressed in Section 5.2.

**Comment:** Want to make sure there is enough money available to address problems that may arise in the future.
**Response:** Injury assessment and restoration planning are ongoing. The Trustees continue to evaluate additional projects for funding as part of the early restoration process but also to work toward developing longer term restoration plans with the goal of fully compensating the public for all resource injuries and losses that resulted from the Spill.

**Comment:** Additional information was provided or suggested for the Trustees to consider when implementing projects.
**Response:** The Trustees appreciate the additional information, which could be useful when considering implementation of additional future restoration projects.

**Comment:** Some commenters expressed general support for the early restoration process and projects proposed in Alternative B.
**Response:** The Trustees acknowledge this support.

**Comment:** Comments suggested other potential restoration projects.
**Response:** The Trustees will continue to evaluate new and existing project ideas as potential DWH NRDA restoration projects. Project ideas can continue to be submitted and reviewed at http://www.gulfspillrestoration.noaa.gov/restoration/.

## Planning and Project Development

**Comment:** Additional transparency in the Trustee DWH oil spill NRDA restoration process was requested.
**Response:** The Trustees understand the importance and value of transparency in the NRDA restoration process and made substantial efforts to ensure the public is aware of the goals of restoration, the criteria to be applied in choosing restoration projects under OPA, the on-going opportunities for the public to submit projects for consideration, and the terms and processes outlined in the Framework Agreement that must also be satisfied to access BP funding. The Trustees have held numerous public meetings and developed and actively manage several web-based information portals used to keep the public apprised about restoration planning for the DWH spill. The Trustees will continue to look for ways to improve their efforts in this regard, provided this can be accomplished consistent with timing, resource, cost and legal constraints.

**Comment:** Consider an ecosystem-based approach to NRDA restoration.
**Response:** The Trustees are mindful of the full array of ecosystem issues in the Gulf region. In undertaking planning for restoration actions, the Trustees have and will continue to consider actions which address ecosystem issues that are consistent with the purposes and goals of restoration under OPA (i.e., compensate the public for losses of natural resources and services resulting from the spill). The Trustees also have and will continue to consider potential impacts on the ecosystem when developing restoration plans.

**Comment:** Explain how Phase II relates to the PEIS and a comprehensive DWH NRDA restoration plan.
**Response:** As this plan is a restoration plan for the NRDA, it has been developed in a manner that allows the Trustees to provide consistency with the anticipated draft PEIS and programmatic restoration plan.

**Comment:** BP should not be allowed to dictate the selection of restoration projects.
**Response:** BP is not dictating the selection of restoration projects. The Trustees are fully responsible for the NRDA for the DWH spill, including all decisions on restoration actions that are appropriate to undertake in compensating for all Spill-related injuries and losses of natural resources and uses in the Gulf. The Framework Agreement makes funding available for Trustee selected projects in return for agreement on Offsets against the Trustees' assessment of natural resource injuries and losses of the public.

## Project Area Contaminants of Concern

**Comment:** Residual DWH oil, response actions and other activities and/or sources of contamination in project areas may negatively impact proposed projects. Coordinate restoration with response (clean-up) activities.

**Response:** Prior to implementing any project the Trustees will coordinate with the Federal On Scene Coordinator to ensure that the project does not obstruct, duplicate or conflict with any ongoing response activities and that any response activities will not obstruct, duplicate or conflict with the project. If such issues arise prior to and/or during project implementation, the Trustees may be able to utilize contingency funds to modify project design, timing and/or otherwise adaptively manage problems.

## Project Implementation

**Comment:** Commenters expressed frustration with the pace of project implementation.
**Response:** The Trustees are working to implement projects as quickly and efficiently as possible following the provisions established in the Framework Agreement. The Framework Agreement outlines a process for the Trustees and BP to reach agreement on the projects to be implemented (after public review and comment), the funding that BP will provide, and the estimated benefits (Offsets) each project provides that would later be credited against the Trustees' total assessment of injury.

**Comment:** The Trustees received several offers for volunteers and partnerships to help with implementation of the projects.
**Response:** The Trustees appreciate offers of partnering and assistance; the implementing Trustees will reach out as needed to make use of available local organizations and resources.

**Comment:** Requests were made for the Trustees to hire local work forces and people negatively impacted by the spill to implement restoration projects.
**Response:** The Trustees support this goal in principle, but recognize that implementing Trustees are subject to and must abide by laws, regulations and policies governing their contracting and procurement processes and practices. Such laws, regulations and policies will vary, depending on the Trustee agency implementing a project. Implementing Trustees will be encouraged to give preference to local hiring to the extent permitted by law.

## Monitoring

**Comment:** Funds should be used to provide for long-term monitoring of fisheries stocks in the Gulf of Mexico.
**Response:** The intent of the early restoration process is to implement projects that accelerate the restoration of resources injured by the Spill. Long-term Gulf monitoring of fisheries, while an important issue, is outside the scope of what the Trustees anticipate accomplishing as early restoration under the terms of the Framework Agreement with BP. The Trustees are continuing to assess the potential injuries and losses to fisheries caused by the Spill and anticipate developing broader monitoring efforts in later stages of the damage assessment and restoration planning process.

**Comment:** It was suggested that early restoration funds be set aside for a long-term Gulf monitoring program, addressing resources and locations beyond the project-specific monitoring efforts identified in the Phase II ERP/ER.

**Response:** The intent of the early restoration process is to implement projects that accelerate the restoration of resources injured by the Spill. Long-term Gulf monitoring, while an important issue, does not meet this objective and is outside the scope of what the Trustees anticipate accomplishing as early restoration under the terms of the Framework Agreement with BP.

**Comment:** Ensure that each project includes project-specific monitoring and success benchmarks to be able to support adaptive management and ensure the public of project success and provide publicly available monitoring information.

**Response:** OPA NRDA regulations set forth several factors that the monitoring component of a Draft Restoration Plan should address to effectively gauge a project's progress and success. Each of the proposed projects in the DERP/ER included a discussion of the performance criteria, monitoring and maintenance plan appropriate for that project. The level of information included is consistent with legal requirements. The Trustees intend to make the results of project activities, including monitoring information, available to the public.

**Other**

**Comment:** People around the world are watching the NRDA restoration process and will hold the Trustees accountable.

**Response:** The Trustees are doing the best job they can to assess the natural resource injuries and compensate the public.

**Comment:** Want to ensure that RESTORE planning and implementation includes public participation and representation by community based groups by those impacted by the Deepwater Horizon spill.

**Response:** The RESTORE Act is outside the scope of the NRDA Early Restoration Process.

**5.2    Comments Specific to Proposed Projects**

**5.2.1    Comments related to both Phase II projects**

**Comment:** How will project provisions be enforced?

**Response:** Trustees acknowledge the importance of enforcement in natural resource conservation. These areas will be subject to law enforcement patrols in accordance with existing practices.

**Comment:** Concern expressed that the funding for these projects could supplant existing funding for similar on-going natural resource management efforts.

**Response:** The Trustees do not intend to supplant or take funding away from existing and ongoing natural resource management efforts. These early restoration projects will enhance existing management efforts.

**Comment:** Clarify which agency(ies) will be responsible for project implementation/success.

**Response:** All Trustees are concerned about project success. The implementing Trustees for Enhanced Management of Avian Breeding Habitat Injured by Response in the Florida Panhandle, Alabama, and Mississippi are FWC, FDEP, and DOI. The implementing Trustees for

Improving Habitat Injured by Spill Response: Restoring the Night Sky are Alabama Department of Conservation and Natural Resources, FWC, FDEP, and DOI.

### 5.2.2 Enhanced Management of Avian Breeding Habitat Injured by Response in the Florida Panhandle

**Comment:** Request that public access to the Gulf Islands National Seashore not be restricted as a result of implementing the avian habitat project.
**Response:** Project implementation will involve only limited periodic public access restrictions to part of the beaches where nesting and foraging activities are being protected in order to achieve the agreed upon improvements. This is consistent with past management actions.

### 5.2.3 Improving Habitat Injured by Spill Response: Restoring the Night Sky

**Comment:** A comment noted the lighting improvements could benefit additional resources, such as migratory birds.
**Response:** This project was designed and selected to address response injury to sea turtle habitat.

**Comment:** Comments that the Night Sky project needs to address/consider implementation on private lands as well as public lands.
**Response:** The Trustees recognize the impacts lighting on private property has on sea turtle habitat, and the goal of education and outreach activities to be undertaken in Florida as a part of this project is to reduce those impacts. As an initial effort, the Trustees have elected to pursue retrofits on the proposed facilities in Alabama and Florida because of the greater degree of certainty and control which helps ensure the anticipated project benefits will be achieved starting with the 2013 nesting season. The Trustees will keep this comment in mind as early restoration proceeds and as other funding mechanisms become available for restoration in the Gulf.

# CHAPTER 6    LITERATURE CITED

Aten, L.E. 1983. Indians of the Upper Texas Coast. Academic Press, New York.

Avissar, N., Burney, C., and Douglass, N. 2012. Guidelines for posting shorebird and seabird sites in Florida. 21 pp.

Brown, C., Andrews, K., Brenner, J., Tunnell, J.W., Canfield, C., Dorsett, C., Driscoll, M., Johnson, E., and S. Kaderka. 2011. Strategy for restoring the Gulf of Mexico (a cooperative NGO report). The Nature Conservancy. Arlington, Virginia. 23 pp.

Coastal Environments, Inc. (CEI). 1982. Sedimentary studies of prehistoric archaeological sites. Prepared for the U.S. Dept. of the Interior, National Park Service, Division of State Plans and Grants, Baton Rouge, LA.

Council on Environmental Quality (CEQ). 1997. Environmental Justice: Guidance under the National Environmental Health Policy Act. Washington, DC: President's Council on Environmental Quality.

Deepwater Horizon Natural Resource Trustees (Trustees). 2012. Deepwater Horizon Oil Spill Phase I Early Restoration Plan and Environmental Assessment. 148 pp.

Gulf Coast Ecosystem Restoration Task Force (GCERTF). 2011. Gulf of Mexico regional ecosystem restoration strategy. 104 pp.

Lohmann, K.J., B.E. Witherington, C.M.F. Lohmann, and M. Salmon. 1997. Orientation, navigation, and natal beach homing in sea turtles. Pages 107-135 in Lutz, P.L. and J.A. Musick (editors). The Biology of Sea Turtles. CRC Press. Boca Raton, Florida.

Mabus, R. 2010. America's Gulf coast: a long term recovery plan after the Deepwater Horizon spill. 130 pp.

Minerals Management Service (MMS). 2007. Gulf of Mexico OCS Oil and Gas Lease Sales: 2007-2012 Final Environmental Impact Statement. U.S. Department of the Interior, Minerals Management Service, Gulf of Mexico OCS Region. http://www.boem.gov/Environmental-Stewardship/Environmental-Assessment/NEPA/nepaprocess.aspx.

National Commission on the BP Deepwater Horizon Oil Spill and Offshore Drilling. 2011. The use of surface and subsea dispersants during the BP Deepwater Horizon oil spill. Staff working paper no. 4. 21 pp. http://www.oilspillcommission.gov/sites/default/files/documents/Updated%20Dispersants%20Working%20Paper.pdf.

National Marine Fisheries Service and U.S. Fish and Wildlife Service (NMSF and FWS). 2008. Recovery Plan for the Northwest Atlantic Population of the Loggerhead Sea Turtle (*Caretta caretta*), Second Revision. National Marine Fisheries Service, Silver Spring, MD.

National Oceanic and Atmospheric Administration (NOAA). 2011. The Gulf of Mexico at a glance: A second glance.
http://stateofthecoast.noaa.gov/NOAAs_Gulf_of_Mexico_at_a_Glance_report.pdf.

National Oceanic and Atmospheric Administration (NOAA), U.S. Department of the Interior, Louisiana Oil Spill Coordinator's Office, Louisiana Department of Environmental Quality, Louisiana Department of Natural Resources, Louisiana Department of Wildlife and Fisheries, 2007. Regional Restoration Plan, Region 2, 24 pp. plus appendices.

Natural Resources Conservation Service (NRCS). 2011. Gulf of Mexico Initiative. U.S. Department of Agriculture. Washington, D.C.

Peterson, C.H., Coleman, F.C., Jackson, J.B.C., Turner R.E., Rowe, G.T., Barber, R.T., Bjorndal, K.A., Carney, R.S., Cowen, R.K., Hoekstra, J.M., Hollibaugh, J.T., Laska, S.B., Luettich, R.A., Osenberg, C.W., Roady, S.E., Senner, S., Teal, J.M., and P. Wang. 2011. A once and future Gulf of Mexico ecosystem: restoration recommendations of an expert working group. Pew Environmental Group, Washington, DC. 112 pp.

Pruner, R.A., M.J. Friel, and J.A. Zimmerman. 2011. Interpreting the influence of habitat management actions on shorebird nesting activity at coastal state parks in the Florida panhandle. 2010-11 study final report. Department of Environmental Protection, Florida Park Service, Panama City, Florida.

Salmon, M. J. Wyneken, E. Fritz, M. Lucas. 1992. Seafinding by hatchling sea turtles: role of brightness, silhouette and beach slope as orientation cues. Behavior: 122(1-2):56-77.

U.S. Army Corps of Engineers (USACE). 2009. Mississippi Coastal Improvements Program, Hancock, Harrison, and Jackson Counties, Mississippi. Comprehensive Plan and Integrated Programmatic Environmental Impact Statement. Volume 1. Main report.

U.S. Army Corps of Engineers (USACE). 2010. The U.S. waterway system: Transportation facts and information. Navigation Data Center.
http://www.ndc.iwr.usace.army.mil/factcard/temp/factcard10.pdf.

U.S. Department of Commerce (USDOC), U.S. Census Bureau. 1999. Poverty.
http://www.census.gov/hhes/www/poverty/.

U.S. Energy Information Administration (USEIA). n.d. Gulf of Mexico fact sheet.
http://www.eia.doe.gov/special/gulf_of_mexico/index.cfm.

U.S. Geological Survey and U.S. Environmental Protection Agency (USGS and EPA). 2011. ESRI maps, National Hydrography Dataset, EPA analyses. Courtesy of Stephen B. Hartley, USGS National Wetlands Research Center.

Witherington, B., P. Kubilis, B. Brost, and A. Meylan. 2009. Decreasing annual nest counts in a globally important loggerhead sea turtle population. Ecol. App. 19(1):30-52.

Witherington, B.E. 1992. Behavioral responses of nesting sea turtles to artificial lighting. Herpetologica 48(1):31-39.

Witherington, B.E. and R.E. Martin. 1996. Understanding, assessing, and resolving light pollution problems on sea turtle nesting beaches. Florida Marine Research Institute Technical Report TR-2. 73 pages.

Witherington, B.E. and R.M. Martin. 2000. Florida Marine Research Institute Technical Reports: Understanding, Assessing, and Resolving Light-Pollution Problems on Sea Turtle Nesting Beaches. Second Edition, Revised. 73 pp.

Appendix A. Federally Listed Threatened and Endangered Species
and Florida Listed Species with the Potential to Occur in
Early Restoration Plan Proposed Project Areas

Table A-1. Species listed by the FWS under the U.S. ESA or by the State of Florida. Note: all federally listed wildlife species in Florida are also listed in Florida.

| Common Name | Species Name | Listing |
|---|---|---|
| American oystercatcher | *Haematopus palliatus* | Florida Species of Special Concern |
| Black skimmer | *Rynchops niger* | Florida Species of Special Concern |
| Brown pelican | *Pelecanus occidentalis* | Florida Species of Special Concern |
| Green sea turtle | *Chelonia mydas* | Federally Endangered |
| Kemps ridley sea turtle | *Lepidochelys kempii* | Federally Endangered |
| Least tern | *Sterna antillarum* | Florida Threatened |
| Leatherback sea turtle | *Dermochelys coriacea* | Federally Endangered |
| Loggerhead sea turtle | *Caretta caretta* | Federally Threatened |
| Piping plover | *Charadrius melodus* | Federally Threatened |
| Snowy plover | *Charadrius nivosus (Charadrius alexandrines)* | Florida Threatened |
| Woodstork | *Mycteria americana* | Federally Endangered |
| St. Andrew beach mouse | *Peromyscus polionotus peninsularis* | Federally Endangered |
| Perdido Key beach mouse | *Peromyscus polionotus trissyllepsis* | Federally Endangered |
| Alabama beach mouse | *Peromyscus polionotus* | Federally Endangered |
| Choctawhatchee beach mouse | *Peromyscus polionotus allophrys* | Federally Endangered |
| Florida perforate cladonia | *Cladonia perforate* | Federally Endangered |

**Appendix B. Potentially
Applicable Laws and Regulations (non-exclusive list)**

1. DOI regulations for implementing NEPA (43 C.F.R. Part 46)
2. Park System Resources Protection Act (16 U.S.C. § 19jj)
3. National Marine Sanctuaries Act (16 U.S.C. §§ 1431 et seq.)
4. Federal Water Pollution Control Act (33 U.S.C. §§ 1251 et seq.)
5. Endangered Species Act (16 U.S.C. §§ 1531 et seq.)
6. NHPA (16 U.S.C. 470 §§ et seq.)
7. Fish and Wildlife Conservation Coordination Act (16 U.S.C. §§ 661-666c)
8. Migratory Bird Treaty Act (16 U.S.C. §§ 703-712)
9. Migratory Bird Conservation Act (126 U.S.C. §§ 715 et seq.)
10. Coastal Zone Management Act (16 U.S.C. §§ 1451-1464)
11. Marine Mammal Protection Act (16 U.S.C. §§ 1361-1421h)
12. Magnuson-Stevens Fishery Conservation and Management Act (16 U.S.C. §§ 1801 et seq.)
13. Clean Air Act (42 U.S.C. §§ 7401 et seq.)
14. Rivers and Harbors Act (33 U.S.C. §§ 401, et seq.)
15. Safe Drinking Water Act (42 U.S.C. §§ 300f et seq.)
16. Noise Control Act (42 U.S.C. §§ 4901 et seq.)
17. Antiquities Act (16 U.S.C. §§ 431 et seq.)
18. Archaeological Resources Protection Act (16 U.S.C. §§ 470aa-470mm)
19. Native American Graves Protection and Repatriation Act (25 U.S.C. §§ 3001 et seq.)
20. Wild and Scenic Rivers Act (16 U.S.C. §§ 1271 et seq.)
21. Historic Sites Act (16 U.S.C. §§ 461-467)
22. Archaeological and Historic Preservation Act (16 U.S.C. §§ 469-469c)
23. Executive Order 11514, Protection and Enhancement of Environmental Quality (Mar. 5, 1970, as amended by Executive Order 11991 (May 24, 1977)
24. Executive Order 11593, Protection and Enhancement of the Cultural Environment (May 13, 1971)
25. Executive Order 11988, Floodplain Management (May 24, 1977)
26. Executive Order 11990, Protection of Wetlands (May 24, 1977)
27. Executive Order 12114, Environmental Effects Abroad of Major Federal Actions (Jan. 4, 1979)
28. Executive Order 12580 (Jan. 23, 1987), as amended by Executive Order 12777, Implementation of Section 311 of the Federal Water Pollution Control Act and the Oil Pollution Act (Oct. 19, 1991)
29. Executive Order 12898, Federal Actions to Address Environmental Justice in Minority Populations and Low-Income Populations (Feb. 11, 1994)
30. Executive Order 12962, Recreational Fisheries (June 7, 1995)
31. Executive Order 13007 – Indian Sacred Sites; and Executive Order 13175 – Consultation and Coordination with Indian Tribal Governments
32. Executive Order 13089, Coral Reef Protection (June 11, 1998)
33. Executive Order 13112, Invasive Species (Feb. 3, 1999)
34. Executive Order 13158, Marine Protected Areas (May 26, 2000)
35. Executive Order 13186, Responsibilities of Federal Agencies to Protect Migratory Birds (Jan. 17, 2001)
36. Executive Order 13352, Facilitation of Cooperative Conservation (Aug. 30, 2004)
37. Subpart G of the National Contingency Plan (40 C.F.R. §§ 300.600 et seq.)

38. White House CEQ regulations for implementing NEPA (40 C.F.R. §§1500 et seq.)
39. DOI Departmental Manual 516 and Environmental Statement Memoranda supplements
40. Anadromous Fish Conservation Act (16 U.S.C. §§ 757[a] et seq.)
41. Coastal Wetlands Planning, Protection and Restoration Act of 1990 (P.L. 101-646)
42. Energy Policy Act (Public Law 109-58, Section 384)
43. Water Resources Development Act (Public Law 110-114, Section 7001-7016)
44. Fish and Wildlife Conservation Act (16 U.S.C. §§ 2901 et seq.)
45. Information Quality Guidelines Issued Pursuant to Section 515 of P.L. 106-554
46. National Wildlife Refuge System Improvement Act of 1997 (16 U.S.C. § 668[dd])
47. Americans with Disabilities Act (P.L. 101-336)
48. Emergency Wetlands Resources Act (16 U.S.C. § 3901)
49. Estuarine Protection Act (16 U.S.C. §§ 1221 et seq.)
50. Marine Protection, Research, and Sanctuaries Act (33 U.S.C. §§ 1401 et seq.)

Appendix C. Acronyms Used in the Early Restoration Plan

APHIS – Animal Plant and Health Inspection Service (USDA)
AR – Administrative Record
BP – BP Exploration and Production, Inc.
CEQ – Council on Environmental Quality
CERCLA – Comprehensive Environmental Response Compensation and Liability Act
C.F.R. – Code of Federal Regulations
CX – Categorical exclusion
DERP – Draft Early Restoration Plan
DOD – Department of Defense
DOI – Department of the Interior
DPEIS – Draft programmatic environmental impact statement
DSAYs – Discounted Service Acre Years
EA – Environmental Assessment
EFH – Essential fish habitat
EIS – Environmental Impact Statement
EPA – Environmental Protection Agency
ER – Environmental Review
ERP – Early Restoration Plan
ESA – Endangered Species Act of 1973
FDEP – Florida Department of Environmental Protection
FONSI – Finding Of No Significant Impact
FPEIS – Final Programmatic Environmental Impact Statement
FWC – Florida Fish and Wildlife Conservation Commission
FWS – Fish and Wildlife Service
GCERTF – Gulf Coast Ecosystem Restoration Task Force
GUIS – Gulf Islands National Seashore
HEA – Habitat Equivalency Analysis
MC252 – Mississippi Canyon 252
NEPA – National Environmental Policy Act
NGMRU – Northern Gulf of Mexico Recovery Unit
NHPA – National Historic Preservation Act of 1966
NOAA – National Oceanic and Atmospheric Administration
NPS – National Park Service
NRDA – Natural Resource Damage Assessment
NWR – National Wildlife Refuge (FWS)
OPA – Oil Pollution Act
PEIS – Programmatic Environmental Impact Statement
RRP – Regional Restoration Planning
SAV – Submerged Aquatic Vegetation
SCAT – Shoreline Cleanup Assessment Team
U.S.C. – United States Code
USDA – United States Department of Agriculture
WS – Wildlife Services (USDA)

Appendix D. Environmental Assessment and Finding of
No Significant Impact for Management of Predation Losses to
State and Federally Endangered, Threatened, and Species of Special
Concern; and Feral Hog Management to Protect Other State and
Federally Endangered, Threatened, Species of Special Concern,
and Candidate Species of Fauna and Flora in the State of Florida,
USDA APHIS WS, 2002

# ENVIRONMENTAL ASSESSMENT

## Management of Predation Losses to State and Federally Endangered, Threatened, and Species of Special Concern; and Feral Hog Management to Protect Other State and Federally Endangered, Threatened, Species of Special Concern, and Candidate Species of Fauna and Flora in the State of Florida

Prepared by:

**UNITED STATES DEPARTMENT OF AGRICULTURE (USDA)**
**ANIMAL AND PLANT HEALTH INSPECTION SERVICE (APHIS)**
**WILDLIFE SERVICES (WS)**

In Cooperation with:

**UNITED STATES DEPARTMENT OF INTERIOR**
U. S. Fish and Wildlife Service
National Park Service

**UNITED STATES DEPARTMENT OF DEFENSE**
U.S. Air Force

**FLORIDA DEPARTMENT OF ENVIRONMENTAL PROTECTION**
Florida Park Service

**FLORIDA FISH AND WILDLIFE CONSERVATION COMMISSION**

## TABLE OF CONTENTS

CHAPTER 3:    ALTERNATIVES

CHAPTER 4:    ENVIRONMENTAL CONSEQUENCES

# ACRONYMS

| | |
|---|---|
| ADC | Animal Damage Control |
| APHIS | Animal and Plant Health Inspection Service |
| AVMA | American Veterinary Medical Association |
| CEQ | Council on Environmental Quality |
| CFR | Code of Federal Regulations |
| DOD | United States Department of Defense |
| EA | Environmental Assessment |
| EIS | Environmental Impact Statement |
| EPA | U.S. Environmental Protection Agency |
| ESA | Endangered Species Act |
| FAC | Florida Administrative Code |
| FDACS | Florida Department of Agriculture and Consumer Services |
| FEIS | Final Environmental Impact Statement |
| FDA | United States Food and Drug Administration |
| FDEP | Florida Department of Environmental Protection |
| FFWCC | Florida Fish and Wildlife Conservation Commission |
| FIFRA | Federal Insecticide, Fungicide, and Rodenticide Act |
| FONSI | Finding Of No Significant Impact |
| FPS | Florida Park Service |
| FS | Florida Statutes |
| FY | Fiscal Year |
| GAO | General Accounting Office |
| IPM | Integrated Pest Management |
| IWDM | Integrated Wildlife Damage Management |
| MBTA | Migratory Bird Treaty Act |
| MOU | Memorandum of Understanding |
| NEPA | National Environmental Policy Act |
| NOA | Notice of Availability |
| NPS | National Park Service |
| NWR | National Wildlife Refuge, U.S. Fish and Wildlife Service |
| ROD | Record of Decision |
| SOP | Standard Operating Procedure |
| T&E | Threatened and Endangered Species |
| USDA | United States Department of Agriculture |
| USDI | United States Department of Interior |
| USFWS | U.S. Fish and Wildlife Service |
| WDM | Wildlife Damage Management |
| WS | Wildlife Services [formerly Animal Damage Control (ADC)] |

## CHAPTER 1: PURPOSE AND NEED FOR ACTION

### INTRODUCTION

Across the United States, natural systems are being substantially altered as human populations expand and encroach on wildlife habitats. Human uses and needs often compete with wildlife for space and resources, increasing the potential for conflicting human/wildlife interactions. In addition, segments of the public strive for protection for all wildlife; this protection can create localized conflicts between humans and wildlife activities. The *Animal Damage Control* (ADC) *Programmatic Final Environmental Impact Statement* (FEIS) summarizes the relationship in American culture of wildlife values and wildlife damage in this way (USDA 1994):

> Wildlife has either positive or negative values, depending on varying human perspectives and circumstances...Wildlife is generally regarded as providing economic, recreational and aesthetic benefits...and the mere knowledge that wildlife exists is a positive benefit to many people. However... the activities of some wildlife may result in economic losses to agriculture and damage to property...Sensitivity to varying perspectives and values are required to manage the balance between human and wildlife needs. In addressing conflicts, wildlife managers must consider not only the needs of those directly affected by wildlife damage but a range of environmental, sociocultural and economic considerations as well.

The United States Department of Agriculture (USDA) is directed by law to protect American agriculture and other resources from damage associated with wildlife. The primary authority for the Animal Damage Control (USDA-Wildlife Services) program is the *Animal Damage Control Act* of March 2, 1931, as amended (46 Stat. 1468; 7 U.S.C. 426-426b and 426c) and the Rural Development, Agriculture and Related Agencies Appropriations Act of 1988 (P.L. 100-202). USDA-Wildlife Services (WS) activities are conducted in cooperation with other federal, state, and local agencies, and private organizations and entities.

Wildlife damage management, or control, is defined as the alleviation of damage or other problems caused by, or related to the presence of wildlife (Leopold 1933, The Wildlife Society 1990, and Berryman 1991). The WS program uses an Integrated Wildlife Damage Management (IWDM) approach (sometimes referred to as IPM or "Integrated Pest Management") in which a series of methods may be used or recommended to reduce wildlife damage. IWDM is described in Chapter 1, 1-7 of the *Animal Damage Control* (ADC) *Programmatic Final Environmental Impact Statement* (USDA 1994). These methods include the alteration of cultural practices as well as habitat and behavioral modification to prevent damage. The control of wildlife damage may also require the removal of an offending animal(s) or the reduction of localized populations of the offending species, through the application of lethal methods. Potential environmental impacts resulting from the application of various wildlife damage reduction techniques are evaluated in this environmental assessment.

According to the Animal and Plant Health Inspection Service procedures implementing the National Environmental Policy Act (NEPA), individual actions are categorically excluded [7 C.F.R. 372.5(c), 60 Fed. Reg. 6,000, 6,003 (1995)]. However, in order to evaluate and determine if there may be any potentially significant or cumulative impacts from the described control program, the Wildlife Services Program in Florida has decided to prepare this environmental assessment (EA).

The purpose of this EA is to analyze the potential effects of the proposed control activities in the State of Florida. This analysis relies predominately on existing federal and state agency publications, information contained in scientific literature, and communications with other wildlife professionals. This EA also

1

cites and is tiered to, the *Animal Damage Control* (ADC) *Programmatic Final Environmental Impact Statement* (USDA 1994).

All control activities will be in compliance with relevant laws, regulations, policies, orders, and procedures, including the Endangered Species Act (ESA). Control activities will not negatively impact other protected flora or fauna. Notice of availability (NOA) of this document will be made consistent with the Agency's NEPA procedures in order to allow interested parties the opportunity to obtain and review this document and comment on the proposed management activities.

## WILDLIFE SERVICES PROGRAM

Wildlife Services (WS) is a cooperatively funded and service oriented program. Before any operational wildlife damage management is conducted, *Agreements for Control* or *WS Work Plans* must be completed by WS and the land owner/administrator. WS cooperates with private property owners and managers and with appropriate natural resource and wildlife management agencies, as requested, with the goal of effectively and efficiently resolving wildlife damage problems in compliance with all applicable federal, state, and local laws and Memorandums of Understanding (MOUs) between WS and other agencies.

Wildlife Services' mission, developed through its strategic planning process, is: 1) to provide leadership in wildlife damage management for the protection of American agriculture, endangered and threatened species, and natural resources, and 2) to safeguard public health and safety. The WS' Policy Manual reflects this mission and provides guidance for engaging in wildlife damage management through:

- close cooperation with other federal and state agencies;
- training of wildlife damage management professionals;
- development and improvement of strategies to reduce losses and threats to publics from wildlife;
- collection, evaluation, and distribution of wildlife damage management information;
- cooperative wildlife damage management programs;
- informing and educating the public on how to reduce wildlife damage and;
- providing data and a source for limited-use management materials and equipment, including federal and state registered pesticides (USDA 1989).

## PURPOSE

In 1998, the U. S. Fish and Wildlife Service (USFWS) sponsored an interagency meeting between State and Federal natural resource managers and the WS to address the need for managing the impacts of predation on endangered and threatened (T&E) species inhabiting Florida's coastal beach and dune ecosystems. The coastal beach and dune ecosystems of Florida support a variety of State and Federally listed species. These species are protected under the Florida and Federal Endangered Species Acts and includes five species of nesting sea turtles, five species of beach mice, one species of cotton mouse, four species of nesting shorebirds, and one species of wintering shorebirds. On April 29, 2000, an additional species was added to this EA, the American crocodile. All agencies represented at this meeting agreed that predation is having a significant impact on the recovery of many of these species. Protection through reduction of predators is necessary to enhance the recovery of these species. The purpose of controlling predation is to maximize chances of survival for these species throughout their coastal ranges. The need for action stems from the low reproductive success, due to documented predation by foxes, raccoons, wild hogs, feral and free-ranging domestic cats, and more recently, coyotes and armadillos.

## PROPOSED ACTION

The WS proposed action for this EA is an Integrated Wildlife Damage Management approach to reduce mammalian predation on T&E species. This alternative would incorporate an integrated management program utilizing certain techniques described in Alternatives 2, 3, and 4 to reduce sea turtle, crocodile, and shorebird nest predation by raccoons, foxes, coyotes, feral hogs, and armadillos; reduce predation threats to beach mice, cotton mice, and adult shorebirds; and reduce predation threats to sea turtle, crocodile, and shorebird hatchlings by raccoons, foxes, coyotes, and feral and free-ranging domestic cats and dogs. This strategy would incorporate non-lethal and lethal control measures.

Management strategies involving exclusion devices would be implemented by natural resource management personnel in accordance with WS recommendations. Local population reduction of predators to reduce immediate predation losses and potential predation threats would be implemented by WS personnel with assistance from the natural resource managers.

## 1.1 NEED FOR ACTION

Humans have brought about the extinction and endangerment of more animals and plants than any other single force of nature, and some contributions leading to extinctions have been caused by the release or escape of domesticated animals (i.e., house cats, dogs, hogs) into newly inhabited environments. Day (1981) addresses at least 9 species of animals that have become extinct as a result of humans, habitat degradation, and the impacts of feral domesticated or imported pests. The following is a synopsis of species whose extinction is believe to have been influenced by European rats, hogs, domestic cats, and dogs: Rodriguez Day Gecko (*Phelsuma edwardnewtoni*; Rodriguez Island); Broad-faced Potoroo (*Potorous platyops*; Western Australia); Gilbert's Potoroo (*Potorous gilberti*; Western Australia); St. Francis Island Potoroo (*Potorous sp.*; St. Francis Island, Australia); Korean Crested Shelduck (*Tadorna (Pseudotadorna) cristata*; Korea); Heath Hen (*Tympanuchus cupido cupido*; New England, USA); Sandwich Rail (*Porzana sandwichensis*; Hawaii, USA); Jamaican Woodrail or Uniform Rail (*Aramides concolor concolor*; Jamaica); and the Dodo (*Raphus cucullatus*; Mauritius Island).

Habitat loss/degradation and other factors have resulted in serious declines in many coastal species throughout their ranges. Habitat loss, storms, predation and other factors have also contributed to serious declines in sea turtles, crocodiles, beach mice, cotton mice, and nesting shorebirds. To compound the threat to endangered and threatened species, some predators have experienced unnatural population increases as a result of human development, elimination of natural predators, ecosystem imbalances, garbage, supplemental feeding, etc. Many T&E species have adapted to very specialized niches and habitats, and are reliant on the few remaining tracts of habitat. In Florida, coastal ecosystems are continually in danger of degradation and influences by humans. T&E species that require this type of habitat generally are more concentrated, and as a result, more susceptible and vulnerable to the effects of heavy predation. This is why protection of T&E species, by reducing predation, is a necessary component in the progression towards their recovery. This EA addresses the need for predator management as it relates to increasing the potential for recovery of these species.

### 1.1.1 Need for Predator Management to Protect Endangered and Threatened Sea Turtles

Five species of sea turtles inhabit the Atlantic and Gulf Coasts of the United States. All are known to nest along the coastal areas of Florida [A. Foley. Florida Department of Environmental Protection (FDEP) pers. comm. Dec. 1998]. The species of concern include: the loggerhead (*Caretta caretta*)

3

(federal; threatened); the green (*Chelonia mydas*) (federal; endangered); the leatherback (*Dermochelys coriacea*) (federal; endangered); the hawksbill (*Eretmochelys imbricata*) (federal; endangered); and the Kemp's ridley (*Lepidochelys kempii*) (federal; endangered). All turtle species listed are protected under the U. S. Endangered Species Act, international agreements, and state laws.

Heavy predation and nest destruction by human activity and a variety of predators have significantly decreased the breeding success of sea turtles. It has been determined that the most significant predators of sea turtle nests are raccoons (*Procyon lotor*), red foxes (*Vulpes* vulpes), coyotes (*Canis latrans*), feral/free-ranging dogs (*Canis familiaris*), feral hogs (*Sus scrofa*), and ghost crabs (*Ocypode* sp.). Recently, in some areas of the southwestern Florida, coyotes have learned to excavate and feed on sea turtle eggs. The nine-banded armadillo (*Dasypus novemcinctus*), has also been observed to excavate and consume sea turtle eggs along some beaches; apparently, this is a new development in armadillo learned behavior. It has become critical for the continued existence of these threatened and endangered sea turtles that nest predation is actively monitored and managed.

Post hatchling predation occurs after hatchlings leave the nest, as they try to make their way to the water. This occurs even when nests are screened to protect against nest predation. Personnel from Eglin Air Force Base have documented hatchlings being preyed upon by coyotes, foxes, raccoons, and ghost crabs after the hatchlings have left the nest. This cannot be controlled except by predator removal.

It is currently estimated, under natural circumstances, that 1 out of 1,000 sea turtle hatchlings survive to breeding age. Responsible natural resource managers seek to increase sea turtle populations by increasing the number of hatchlings that reach the sea. As suitable nesting habitat dwindles it will be essential that nest production be maximized in productive nesting areas. This can only be accomplished through the direct management of predators inhabiting areas critical to the survival of these T&E species.

The FDEP-Florida Park Service (FPS) suggests that the State's overall sea turtle nesting success may fluctuate around 55% depending on weather, predation, and other factors per given year. In 1998, 74% of the State and Federal natural resource managers in the Florida panhandle reported predation on sea turtle nests. Predation on sea turtle nests has becoming a more significant concern to resource managers statewide. Natural resource managers also acknowledge that some areas of the state may experience little to no nest predation, while others experience heavy losses.

Prior to 1998, FDEP authorized some permit holders to initiate wildlife damage management efforts to alleviate nest predation. In 1998 sea turtle permitting and management efforts were transferred to the Florida Fish and Wildlife Conservation Commission (FFWCC). These efforts include placing wire excluders over turtle nests to prevent coyotes and other predators from excavating the eggs. Unfortunately, the management efforts currently employed by many permit holders have not significantly reduced nest predation. Reasons for this limited success include: predators actively patrol the beaches at night and raid nests prior to the placement of wire excluders; the topography and sandy soil of the coastal dune regions limit accessibility to many nesting areas; the use of ineffective predator control techniques; and many predators have learned to by-pass excluding devices. Not all predators have learned to dig under excluders; therefore, in many cases, only a few animals represent a significant problem. However, it is believed that this new behavior is learned and has the potential to be passed on to other individuals in the area. This being the case, it is of critical importance to selectively remove individual predators that are by-passing the excluding devices and actively preying on turtle eggs.

Predator density often is limited by suitable habitat and the availability of other essential resources. Coastal habitats may differ considerably between regions of the state. As a result, not all natural resource managers will experience the same type or abundance of predators throughout the state. For

4

example, raccoons have been documented as the major nest predator in south Florida; whereas, coyotes and foxes have been documented as the major nest predators in northwest Florida.

## Coyotes

The presence of coyotes in Florida is thought to be the result of human introductions of western coyotes during the 1920's and range expansion of populations from adjoining states (Bekoff 1977, Cunningham and Dunford 1970, Paradiso 1968). Coyotes are known to have been well established in the panhandle and north-central Florida regions of the state for many years, and the coyote is now believed to occur throughout most of the Florida peninsula. The coyote is expected to continue its range expansion throughout the remainder of the State (Parker 1995).

In the last decade, coyotes have become the most efficient predator of sea turtle nests in northwestern Florida. FPS biologists have regularly monitored sea turtle nesting activity in the panhandle region for decades and began noticing nest predation by coyotes in the early 1990's. Since then, the FPS has documented coyote nest predation and has found this type of predation to be significant to the nesting success of sea turtles in many areas of the northwest Florida.

Since 1993, documented predation by coyotes of sea turtle nests in the St. Joseph Peninsula State Park increased from 43.2% in 1994 (36 of 88 nests), to 52.8% in 1996 (47 of 89 nests). Late in the 1995 nesting season, coyotes successfully predated sea turtle nests protected by excluders. In 1995, nest predation averaged one nest per night until Hurricane Opal destroyed all of the remaining sea turtle nests. In 1997, in a cooperative effort with St. Joseph Peninsula State Park, the U.S. Fish and Wildlife Service (USFWS) entered into an agreement with Wildlife Services to initiate an integrated wildlife damage management plan for the St. Joseph Peninsula State Park to reduce predation on sea turtle nests, and to reduce coyote predation on the St. Andrews beach mouse. As a result of this management effort, nest predation was reduced to 6.3% (8 of 126 nests); predation was reduced by 88% from the previous year.

Gulf Islands National Seashore (GINS) on Perdido Key, in northwestern Florida, also experienced heavy predation on sea turtle nests. In 1997, 70 % of all sea turtle nests were lost to coyote and red fox predation. In the spring of 1998, three of the first four turtle nests of the season were predated. At the request of the USFWS and GINS, WS implemented an emergency wildlife management plan encompassing an eight mile section of Perdido Key. Predation stopped after one coyote and five foxes were removed.

Eglin Air Force Base has recently experienced heavy predation losses of sea turtle nests by coyotes, foxes, and raccoons. In spite of the installation of wire excluders on sea turtle nests, predation rates were 62% (26 of 42 nests) in 1996 and 61 % ( 14 of 23 nests) in 1997. In 1998, the USFWS and Eglin's Natural resource managers requested WS assistance in implementing an emergency IWDM plan. Prior to the implementation of the IWDM plan, 60 % of the existing nests (9 of 15 nests) had experienced depredation. After implementation of the plan the percentage of new nests depredated in 1998 dropped to 17 % (3 of 17 nests). In 1999 on Eglin's restricted Santa Rosa Island beach (13 miles), all nests that were not screened were destroyed by predators (9 of 15; totaling 60%).

## Raccoons

Raccoons are by far the most abundant native predator in Florida. The FDEP estimates that 90% of all reported sea turtle nest predation in south Florida is caused by raccoons. In 1996, at a sea turtle seminar in Jensen Beach, Florida, it was the consensus amongst sea turtle biologists that raccoon predation represents one of the most significant threats to sea turtle nesting in the Americas. Some of the reasons for this threat is the fact that raccoons have relatively few enemies, are extremely adaptable, and have relative high populations throughout much of their range.

In a publication released by the National Academy of Sciences (1990), raccoons were considered the most significant predator of loggerhead turtle eggs in the Southeast. An excerpt from this publication describes the role raccoons play in sea turtle nest predation:

> The major loggerhead egg predator in the southeastern United States is the raccoon (Dodd, 1988).
> Before protective efforts were initiated, raccoons destroyed nearly all the nests at Canaveral National Seashore, Florida (Ehrhart, 1979), and at Cape Sable, Florida, raccoons destroyed 85% of the nests in 1972 and 75% in 1973 (Davis and Whiting, 1977). The High rate of predation might have resulted from the unusually large raccoon populations, which were augmented by such human activities as habitat alteration, food supplements (garbage), and removal of natural predators of the raccoon (Carr, 1973; pers. comm., L. Ehrhart, University of Central Florida, 1989). Not all nesting beaches in Florida suffer such high losses from raccoons; for example, only seven of 97 nests on Melbourne Beach, Florida, were destroyed by raccoons in 1985 (Witherington, 1986). Other nest predators are ghost crabs, hogs, foxes, fish crows, and ants (Dodd, 1988). From 1980 to 1982, nonhuman predators destroyed up to 80% of the loggerhead clutches laid on two barrier islands in South Carolina (Hopkins and Murphy, 1983).

Predation rates at Hobe Sound NWR, in southeast Florida, were as high as 95% prior to predator management activities. During the 1972-1977 sea turtle nesting seasons, raccoons were trapped and removed from Hobe Sound NWR. This activity reduced nest predation to under 6% during those years. In 1978, trapping activity reduced losses to under 2% of pre-trapping predation rates ( 11 of 969 nests were lost). During this same period, predation losses in an untrapped 2-mile stretch of beach, on St. Lucie Inlet State Park immediately north of the Hobe Sound NWR boundary, were over 50%.

Raccoons have also been document to be the most important predator of sea turtle eggs at Ten Thousand Islands National Wildlife Refuge (TTINWR), in extreme southwest Florida. Raccoon predation was determined to range from 49-87% between 1991-1994 at the TTINWR. As a result of this high predation rate, the USFWS contracted the University of Florida to conduct a research project to determine the effects of raccoon trapping as a means to reduce raccoon predation on sea turtle nests on 4 islands within the Ten Thousand Islands area. One island in particular, Panther Key (54.8 ha), was selected for control work because of the fairly extensive pretrapping data that existed for this island since 1991. In 1995, the research project was started on these islands. A total of 14 raccoons were removed from Panther Key during the 1995 season and nest surveys showed a significant decrease in nest predation (Table 1-1). However, since 1996 maintenance trapping efforts have been limited and have resulted in a steady increase in sea turtle nest predation by raccoons (Garmestani 1997, Tamalis and Doyle 1999).

Table 1-1. Nine years of sea turtle survey information for the Panther Key Study Site. No raccoon predation was documented or observed following raccoon removals from the island in 1995.

| YEAR | Total Nests | # Predated | % Predated |
|---|---|---|---|
| 1991 | 72 | 63 | 87.5 |
| 1992 | 42 | 40 | 95.2 |
| 1993 | 28 | 20 | 71.4 |
| 1994 | 42 | 29 | 69 |
| 1995* | 41 | 0 | 0 |
| 1996 | 62 | 2 | 3.2 |
| 1997 | 94 | 32 | 34 |
| 1998 | 61 | 42 | 68.9 |
| 1999 | 80 | 27 | 33.8 |

* Year in which intense raccoon trapping was conducted.

A study conducted in the Everglades National Park, reported raccoon predation on 75-85% of loggerhead sea turtle nests in one area (Davis and Whiting 1977). Raccoon control on this same beach reduced predation by 46%. Johnson and Rauber (1970) found that raccoon control on the Cape Romain National Wildlife Refuge decreased loggerhead sea turtle nest predation from ~ 80% to 2%.

## Armadillos

In the past few years, Hobe Sound NWR personnel have documented non-native armadillos digging into sea turtle nests and feeding on the eggs (R. M. Noel. USFWS. Hobe Sound National Wildlife Refuge. pers. comm. February 2000). This may seem odd when most research indicates that the diet of armadillos generally consists of insects, other arthropods, and small vertebrates (i.e., salamanders, lizards, etc.); however, there have been numerous accounts of armadillos feeding on ground nesting bird, reptile, and amphibian eggs as well. It is also conceivable that armadillos have learned to excavate and feed on the eggs of sea turtles in some areas of Florida.

## Feral (Wild) Hogs

Hogs were introduced to Florida by the Spanish explorer Hernando de Soto in 1539. Florida has the second largest number of wild hogs in the United States, second only to Texas. Wild hogs are found in all 67 counties in Florida and are considered game animals on 45 Wildlife Management Areas, 2 Wildlife and Environmental Areas, and in parts of Collier, Dade, and Monroe counties. On these areas wild hogs are protected by state law. On other lands in Florida, hogs are classified as domestic livestock and are the property of the landowner.

Feral hogs are known nest predators of sea turtles throughout their range [i.e., Southeast United States, Galapagos Islands, Mexico, Costa Rica, Australia, Tortuguero (Stancyk 1979)]. Many state and federal natural resource managers are now in the process of controlling hog numbers because of their known impact to endangered plants and animals (Thompson 1977). Feral hogs are not native to North America and many native species have not evolved to deal with hog competition or predation. Feral hogs are known to feed on many of the smaller animals (some threatened or endangered), disrupt ecosystems via rooting, and feeding on rare and endangered plants.

Natural resource agencies report that non-native hogs have destroyed up to 80% of endangered sea turtles nests in some undeveloped coastal regions of Florida. Cape Canaveral, St. Vincent NWR's, and Cape St. George Island are three other areas where wild hogs have been documented to actively predate on sea turtle nests. Some federal and state officials have introduced management actions to help control feral hog populations on federal and state lands.

## Feral/Free-Ranging Cats and Dogs

There appears to be some discrepancy between both wildlife professionals and lay persons as to what constitutes a feral animal. Van't Woudt (1990) uses three categories to classify the status of a domesticated animal observed in the wild: 1) an animal that stays in close proximity to its home or owner (tame); 2) an animal that may or may not have a home or owner but is reliant on humans for shelter and food (free-ranging); and 3) an animal that breeds and lives without human interactions (feral). For the purpose and scope of this EA, the Florida WS Program will adopt Van't Woudt's (1990) definitions of tame, free-ranging, and feral domesticated animals, as described above. Additionally, WS will consider all domesticated species or breeds as feral or free-ranging animals when captured during control operations, unless an animal is readily identified with a collar and/or an identification tag.

Domesticated cats (*Felis catus*) and dogs have been identified as significant nest and/or hatchling predators of sea turtles. A study in Aldabra Atoll, Seychelles, found feral cat predation to have a significant impact on green turtle hatchlings. Seabrook (1989) found a positive correlation in cat activity and green turtle nesting at Aldabra Atoll (r=646, d.f.=21, P<0.001). In a survey of reported predators of sea turtle nests and hatchlings, Stancyk (1979) found feral and free-ranging dogs to be significant predators in the Galapagos Islands, Tortuguero, South Africa, Mexico, and South Yemen.

## 1.1.2 Need for Predator Management to Protect Endangered and Threatened Beach Mice, Cotton Mice, Woodrats, Rice Rats, & Lower Keys Marsh Rabbits

Seven extant subspecies of beach mice inhabit the Atlantic and Gulf Coasts of the United States. Six federally listed endangered or threatened species of mice, two species of endangered rats, and one species of endangered rabbit are found along the Florida's coastal regions and include the following: Perdido Key beach mouse (*Peromyscus polionotus* trissyllepsis) (federal; endangered); Saint Andrews beach mouse (*Peromyscus polionotus peninsularis*) (federal; endangered); Anastasia Island beach mouse (*Peromyscus polionotus phasma*) (federal; endangered); Choctawhatchee beach mouse (*Peromyscus polionotus allophrys*) (federal; endangered); Key Largo cotton mouse (*Peromyscus gossypinus allapaticola*) (federal; endangered); Key Largo Woodrat (*Neotoma floridana* smalli) (federal; endangered); Southeastern beach mouse (*Peromyscus polionotus niveiventris* ) (federal; threatened); silver rice rat (*Oryzomys palustris natator*) (federal; endangered); and the Lower Keys marsh rabbit (*Sylvilagus palustris hefneri*) (federal; endangered) . An additional species, the Santa Rosa beach mouse (*Peromyscus polionotus leucocephlus*) is listed as a Species of Special Concern. The suspected and potential predators of these endangered mammals include feral/free-ranging house cats, bobcats (*Felis rufus*), foxes, coyotes, feral/free-ranging dogs, black rats (*Rattus* rattus), raccoons, skunks (*Mephitis mephitis* and *Spilogale putorius*), armadillos, owls (Tytonidae and Strigidae), hawks (Accipitridae), great blue herons (*Ardea herodias*), snakes (*Masticophis flagellum, Coluber constrictor, and Elaphe spp.*) and red-imported fire ants [(*Solenopsis sp.*) USFWS 1999].

### Feral/Free-Ranging Cats & Dogs, Black Rats, Feral Hogs, Foxes, and Coyotes

In 1995, the USFWS contracted Auburn University to conduct a 3-year beach mouse survey. During the survey, low trapping success was documented in areas where house cat tracks were observed. Cat tracks have been documented in all environs of the beach mouse. Feral and free-ranging domestic cats have a documented higher abundance in critical beach mouse habitat located in close proximity to urban development (Moyers 1996, C. Petrick, Eglin AFB, pers. comm., Dec.1998).
A small number of Perdido Key beach mice, estimated ≤ 100 (M. Wooten, Auburn University, pers. comm. Dec. 1998), is the only known extant population. Losses have been attributed to natural disasters (i.e., hurricanes, erosion of shorelines, etc.), habitat losses (i.e., land development), and predation. Biologists are concerned that without intensive management, including predator control, this subspecies will soon become extinct. Predation appears to be a significant factor contributing to the demise of this beach mouse. Feral cats, foxes, and coyotes have been documented as major predators of the beach mouse on Gulf Islands National Seashore (GINS). Florida Park Service biologists at Perdido Key have noted an increased number of these predators at the Perdido Key State Recreation Area. In the past, Park managers have attempted to control the increasing predator population without success. Recently, the USFWS requested the WS to assist the Florida Park Service in controlling beach mouse predation.

The Choctawhatchee beach mouse inhabits Shell Island in northwest Florida. In 1998, Hurricane George reduced mouse populations to critically low levels, and biologists are concerned that this subspecies may be extirpated. Controlling predation by feral and free-ranging domestic cats could be a critical factor in saving the Choctawhatchee beach mouse from extinction.

The Santa Rosa beach mouse inhabits an undeveloped section of Santa Rosa Island. Wildlife biologists at Eglin Air Force Base report that feral and free-ranging domestic cats, and possibly foxes, threaten the stability of the Santa Rosa beach mouse. The high abundance of feral and free-ranging domestic cats on Santa Rosa Island has caused great concern for federal natural resource managers and regulators about the stability of the population on the island. Management efforts are underway to assure stability and increase of the population and, since feral and free-ranging domestic cats are believed to be major predators of the beach mouse, controlling cats must be a part of those efforts.

A viable population of the St. Andrews beach mouse inhabits Saint Joseph Peninsula State Park. While the extent of predation on the St. Andrews beach mouse is not fully known, biologists from the USFWS have expressed concern about the potential impacts of coyote predation. USFWS has identified the viability of this population as essential to future recovery efforts. This subspecies was extirpated from the Tyndall Air Force Base; in 1998, the St. Andrews beach mouse was reintroduced on the Tyndall AFB (J. E. Moyers, Auburn University; pers. comm., Dec.1998).

The Anastasia Island beach mouse is one of only two subspecies inhabiting the Atlantic coast of Florida. The historical range of this subspecies of beach mouse extended from the Duval-St. Johns County line to Matanzas Inlet, St. Johns County, Florida (roughly, 50 linear miles). Currently, this subspecies inhabits approximately three miles of beach/dune habitat on Anastasia Island. Both federal and state biologists have strong concerns about increased human development and the potential of feral/free-ranging cat and dog predation on beach mice in these areas. Biologists are also concerned about potential house mouse and rat competition with the native beach mouse along these developed areas.

The southeastern beach mouse is the second subspecies found on the Atlantic coast of Florida. Its historical range extended from Ponce Inlet, Volusia County to Miami Beach in Dade County, Florida ($\simeq$ 175 linear miles. Currently, this mouse occupies only 50 miles of its previous range, predominately on federal, state, and county owned lands. Both federal and state biologists have strong concerns about increased human development and the potential of feral/free-ranging cat and dog predation on beach mice in these areas. Biologists are also concerned about potential house mouse and rat competition with the native beach mouse along these developed areas.

The Key Largo cotton mouse and wood rat are endemic rodents to Key Largo. The only known populations of these two endangered rodents are restricted to the northernmost portion of this Key. The USFWS and other conservation agencies are concerned about the effects feral/free-ranging dogs and cats, black rats, and raccoons will have on the recovery efforts of these species (USFWS 1999). Currently, WS is not aware of any control measures that are being implemented to manage and/or reduce predation and competition threats from the above listed species.

The silver rice rat and the Lower Keys marsh rabbit are two endemic mammal species restricted to the Lower Keys Region. Both of these endangered mammals are found in the coastal marshes and wetlands of this area and share these habitats with other endangered animals, including nesting Atlantic loggerhead and green sea turtles. Recovery biologists are concerned with all aspects of the recovery of these species including predation and competition from free-ranging dogs and cats, black rats, and raccoons (USFWS 1999).

9

There are no known cases where feral hogs have been observed to root up and feed on beach mice or other endangered mammals. The problem with beach mouse/feral hog interactions is the competition for food resources and habitat destruction. It has been well documented that feral hogs disturb large areas of vegetation and soil through rooting, and it is suspected that hogs inhabiting coastal ecosystems are uprooting and damaging vegetation considered essential for beach mouse winter foods [i.e., sea oats (*Uniola paniculata*), beach grass (*Panicum spp.*), blue stem (*Schizachyrium* maritimum), beach pea (*Galactia sp.*)] and dune stabilization. It has been documented that hogs can disrupt natural vegetative communities, eliminate rare plants and animals, and promote the expansion of exotic plant species by soils disturbance.

### 1.1.3 Need for Predator Management to Protect Endangered and Threatened Shorebirds and Other Listed Colonial Nesting Bird Species

There are five species of colonial and/or shore-nesting bird species that nest in the sand dune and interdunal habitats along Florida's coastline that are listed as threatened or species of special concern, and one species that winters along Florida's coasts. Listed shore-nesting species in Florida include the following: roseate tern, *Sterna dougallii dougallii* (federal; threatened); southeastern snowy plover, *Charadrius alexandrinus tenuirostris* (state; threatened); American oystercatcher, *Heamatopus palliatus* (state; species of special concern); black skimmer, *Rynchops niger* (state; species of special concern); and least tern, *Sterna antillarum* (state; threatened). The one Listed species of shorebird that only winters in Florida is the piping plover, *Charadrius melodus* (federal; threatened).

Populations of shore-nesting birds flourished on the Gulf Islands National Seashore (GINS) in the 1970's. Nesting species included oystercatchers, black skimmers, least terns, and southeastern snowy plovers. In addition to habitat degradation, predation by red foxes, coyotes, and feral cats on GINS has contributed to the decline in its nesting shorebirds. Historically, several thousand pairs of shorebirds nested at GINS; only 15 pairs were documented in 1998. The southeastern snowy plover is the only species currently nesting on GINS, and nest predation has significantly affected hatching success.

The Caribbean subspecies of roseate tern is listed as threatened in the United States and is known to nest only in the Dade and Monroe counties of Florida. Roseate terns are colonial nesters and often nest in association with least terns on beach habitats and on some rooftops in Florida. Throughout their range, roseate tern colonies have a multitude of predators that include birds, mammals, and invertebrates. Mammalian predators that are of concern in the Florida Keys are raccoons, rats, and potentially feral/free-ranging cats (USFWS 1999).

**Feral/Free-Ranging Cats & Dogs, Feral Hogs, & Other Documented/Suspected Predators**

The seventeen mile long beach at Eglin Air Force Base (AFB) on Santa Rosa Island provides prime undeveloped coastal beach habitat. A recent study of southeastern snowy plover nesting sites, from Texas to south Florida, suggests that 53% of the total population nests on Eglin's sea shore (C. Petrick; pers. comm.; Dec.1998). Currently, as a result of predation (e.g., coyotes, foxes, raccoons, feral and free-ranging domestic cats), Eglin does not have any significant colonial shorebird nesting sites. Eglin does have a significant population of solitary nesting snowy plovers; consequently, snowy plover nests are more spatially dispersed, making them less vulnerable to the levels of predation incurred by colonial nesting species. Feral cats are a major concern, and population reduction efforts of the feral cats are being conducted.

Massey (1971) and Massey and Atwood (1981) found that predators can prevent least terns from nesting or cause them to abandon previously occupied sites. In another study, mammalian predators were found to have significantly impacted the loss of least tern eggs on sandbars and sandpits (Kirsch 1996). Skunks (Massey and Atwood 1979), red foxes (Minsky 1980), coyotes (Grover and Knopf 1982), and raccoons (Gore and Kinnison 1991) are common predators of least terns. During one 2-year study, coyotes destroyed 25.0-38.5% of all interior least tern nests (Grover 1979). Raccoons are considered a major predator of ground-nesting upland bird nests and poults (Johnson 1970, Speake 1980, Speake et al. 1985, Speake et al. 1969).

In Massachusetts, predators destroyed 52-81% of all active piping plover nests from 1985-1987 (Macro et al. 1990). Red foxes accounted for 71-100% of the nests destroyed by predators at the site. During FY95-98, Nebraska personnel were asked to remove coyotes, striped skunks, opossums *(Didelphis virginiana)*, and mink *(Mustela vison)* from nesting sites along the Platte River in central Nebraska to protect threatened piping plovers and endangered least terns. As expected, the removal of predators increased plover and tern nesting success and chick survival rates (Wildlife Services 1999.)

Balser et al. (1968) recommended that predator damage management programs target the entire predator complex or compensatory predation may occur by a species not under control, a phenomena also observed by Greenwood (1986). Trautman et al. (1974) concluded that a single species predator damage management program showed some promise for enhancing ring-necked pheasant *(Phasianus colchicus)* populations. Clearly, predator damage management can be an important tool for achieving and maintaining game, nongame, and T&E species production and management objectives.

The Florida Fish and Wildlife Conservation Commission (1999) regularly monitors breeding colonies of known colonial shorebirds in Florida. Eighty-seven nesting colonies were monitored and data were collected on the predation within these colonies for 1998-1999. Of the 87 colonies, 32 showed signs of possible predation from various predator species. Ten species or species-groups of predators were documented at these colonies and include the following: feral cats, dogs, raccoons, laughing gulls *(Larus atricilla)*, crows *(Corvus spp.)*, herons, feral hogs, grackles *(Quiscalus spp.)*, coyotes, and bobcats. Shorebird species incurring the greatest predation were least terns, laughing gulls, and black skimmers. Data indicate that raccoons, crows, and feral cats were the most significant predators of shorebird colonies (Figure 1-1). Mammalian predators account for 63% of the total suspected predation on colonial shorebirds nesting in Florida. Of the 63%, raccoons and feral/free-ranging domesticated species accounted for more than 90% of the suspected predation to shorebirds by mammals, for 1998-99.

Figure 1-1. Colonial shorebird breeding colonies in Florida. Suspected predation of colonial shorebird nesting sites and the predators involved.

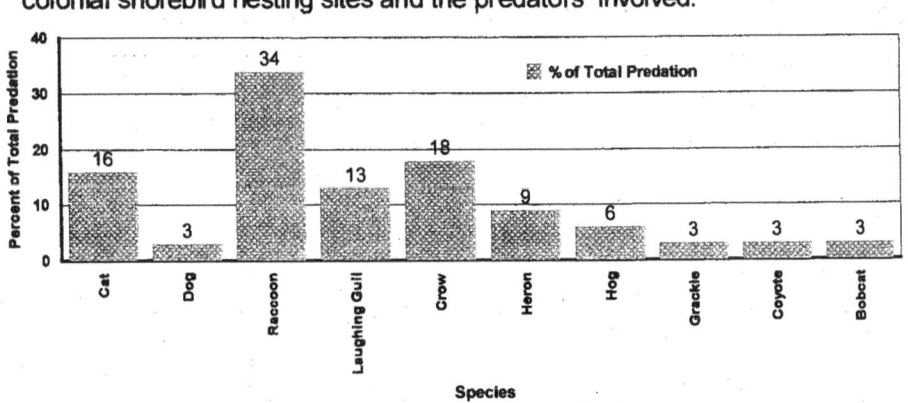

### 1.1.4 Need for Predator Management to Protect Endangered American Crocodile Nests

The American crocodile (*Crocodylus acutus*) is one of two species of crocodilians native to Florida; the second species is the American alligator *(Alligator mississippiensis)*. The American crocodile is restricted to wetland and mangrove habitats of south Florida and overlap very little with the alligator. American crocodiles were listed as an endangered species in 1979 by the USFWS and recovery efforts for the species were prioritized. In most areas of crocodile habitat in south Florida, nest predation has not been a limiting factor (USFWS 1999). However, crocodile nests located in areas of high raccoon densities (i.e., Cape Sable) have been observed to suffer exceedingly high damage from this predator (S. Snow. NPS. Everglades National Park. pers. comm. March 2000). Raccoon nest predation appears to be localized and restricted to areas of high raccoon densities. Currently, all other nest predators are not considered a significant threat to the local or regional recovery potential of crocodiles.

### 1.1.5 Need for Feral Hog Management to Protect State and Federally Endangered, Threatened, Species of Special Concern, and Candidate Species of Fauna and Flora

Many experts in the fields of botany and herpetology have observed marked declines in some rare species of plants, reptiles, amphibians, and soil invertebrates (Singer et al. 1982) in areas inhabited by feral hogs (or wild hogs). It has been well documented that feral hogs disturb large areas of vegetation and soils through rooting, and it is documented that hogs inhabiting coastal, upland, and wetland ecosystems are uprooting, damaging, and feeding on rare native species of plants and animals (Means 1999). It has been documented that hogs can disrupt natural vegetative communities, eliminate rare plants and animals, alter species composition within a forest [both canopy and low growing species (Frost 1993, Lipscomb 1989)], increase water turbidity in streams and wetlands (reducing water quality and impacting native fishes), increase soil erosion and alter nutrient cycling (DeBenedetti 1986, Singer et al. 1982), and promote the expansion of exotic plant species by soil disturbance (Southwest Florida Water Management District 1996).

Nearly twenty-two plant species and four species of amphibians listed as rare, threatened, endangered, or species of special concern have been affected by feral hog activities at the Eglin Air Force Base. Many of these species inhabit habitats that are themselves becoming rare and threatened by human uses [i.e., seepage bogs, flatwoods, wet prairies, floodplain forests, sandhill communities, etc. (Printiss and Hipes 1999)]. Florida Natural Areas Inventories, conducted by the Nature Conservancy, implicate feral hogs as a major negative influence of native systems in Florida and recommends that hog management be a major focus for natural resource managers with conservation minded programs.

The following is a list of animals and plants that are considered to be threatened by hog activities on the Eglin Air Force Base, Florida: flatwoods salamander, *Ambystoma cingulatum* (federal; threatened); gopher frog, *Rana areolata* (federal; C2); bog frog, *Rana okalossae* (federal; C2); dwarf salamander, *Eurycea quadridigitata* (federal; C2); Chapman's aster, *Aster chapmanii* (federal; C2); coyote-thistle aster, *Aster eryngiifolius* (federal; C2); Curtiss' sand grass, *Calamovilfa curtissii* (federal; C2); water sundew, *Drosera intermedia* (state; threatened); Florida anise, *Illicium floridanum* (state; threatened); bogbuttons, *Lachnocaulon digynum* (federal; C2); Catesby's lily, *Lilium catesbaei* (state; threatened); panhandle lily, *Lilium iridollae* (federal; C2); West's flax, *Linum westii* (federal; C2); west Florida cow lily, *Nuphar luteum ulvaceum* (federal; C2); naked-stemmed panic grass, *Panicum nudicaule* (federal; C2); Chapman's butterwort, *Pinguicula planifolia* (federal; C2); butterwort - unnamed, *Pinguicula primuliflora* (state; threatened); southern yellow fringeless orchid, *Platanthera integra* (state; threatened); willow-leaved meadowbeauty, *Rhexia salicifolia* (federal; C2); Alabama beakrush, *Rhynchospora crinipes* (federal; C2); white-top pitcher plant, *Sarracenia leucophylla* (federal; C2); parrot pitcher plant, *Sarracenia psittacina* (state; threatened); sweet pitcher plant, *Sarracenia rubra* (state; endangered); Drummond's yellow-eyed grass, *Xyris drummondii* (federal; C2); karst pond yellow-eyed grass, *Xyris longisspala* (state; endangered); and Harper's yellow-eyed grass, *Xyris scabrifolia* (state; threatened).

## 1.2    FLORIDA WILDLIFE SERVICES OBJECTIVES

The need to manage predator impacts on endangered, threatened, and species of special concern was used by WS, with input from the USFWS, NPS, FDEP, FFWCC, and the DOD (U. S. Department of Defense), to define the objectives for the WS program in Florida. Florida WS' objectives for the protection of endangered and threatened species along the coastal habitats of Florida and for cooperative agreements and agreements for control within the State are to:

- Respond to 100% of the requests for assistance with the appropriate action (technical assistance or direct control) as determined by Florida WS personnel, applying the ADC Decision Model (Slate et al. 1992).

- Hold sea turtle nest predation to less than 20% per year, on properties with a federal WS operational program.

- Hold American crocodile nest predation to less than 20% per year, on properties with a federal WS operational program.

- Hold beach mouse and nesting-wintering shorebird predation to less than 20% per year, on properties with a federal WS operational program.

- Reduce feral hog populations to the greatest extent possible, on properties with a federal WS operational program.

- Maintain the lethal take of nontarget animals by WS personnel during damage management to less than 10% of the total animals taken.

## 1.3    RELATIONSHIP TO OTHER ENVIRONMENTAL DOCUMENTS

**ADC Programmatic EIS.** WS [formerly known as Animal Damage Control (ADC)] has issued a Final Environmental Impact Statement (FEIS) on the National APHIS/WS program (USDA 1994). Pertinent and current information available in the Final EIS has been incorporated by reference into this EA.

## 1.4    DECISION TO BE MADE

Based on agency relationships, MOUs, and legislative authorities, Florida WS is the lead agency for this EA, and therefore, is responsible for the scope, content, and decisions made. The USFWS,

NPS, DOD, FDEP, and the FFWCC provided input throughout the EA preparation process to ensure an interdisciplinary approach according to NEPA and agency mandates, policies, and regulations.

Based on the scope of this EA, the decisions to be made are:

- Should predator damage to T&E species be allowed to continue without a WS predator management program?

- If so, how should WS fulfill its legal responsibilities to protect T&E species in Florida?

- Would the proposed action have significant impacts requiring an EIS analysis?

## 1.5   SCOPE OF THIS EA ANALYSIS

**Actions Analyzed.**  This EA evaluates planned predator damage management to protect endangered, threatened, and species of special concern in the state of Florida from mammalian predators.  Additional NEPA documentation would be required to conduct wildlife damage management that is outside the scope of this EA, should the need arise.

**Wildlife and Plant Species Potentially Protected by Florida Wildlife Services.**  The USFWS, NPS, DOD, FDEP, FFWCC, or other entities may request Florida WS assistance to achieve management objectives for the loggerhead, green, leatherback, hawksbill, and Kemp's ridley sea turtles;  American crocodile; the Perdido Key beach mouse, St. Andrews beach mouse, Choctawhatchee beach mouse, Anastasia Island beach mouse, Southeastern beach mouse, Key Largo cotton mouse, Key Largo woodrat, silver rice rat, Lower Keys rabbit; and the roseate tern, southeastern snowy plover, piping plover, American oystercatcher, black skimmer, and the least tern.

Additional plant and animal species that would benefit from feral hog control include: flatwoods salamander, gopher frog, bog frog, dwarf salamander; and Chapman's aster, coyote-thistle aster, Curtiss' sand grass, water sundew, Florida anise, bogbuttons, Catesby's lily, panhandle lily, West's flax, west Florida cow lily, naked-stemmed panic grass, Chapman's butterwort, butterwort - unnamed, southern yellow fringeless orchid, willow-leaved meadowbeauty, Alabama beakrush, white-top pitcher plant, parrot pitcher plant, sweet pitcher plant, Drummond's yellow-eyed grass, karst pond yellow-eyed grass, and Harper's yellow-eyed.

If other species are identified as in need of protection from predators or feral hogs, a determination regarding the need for additional NEPA analysis would be made on a case-by-case basis.

**Period for Which this EA is Valid.**  This EA would remain valid until Florida WS and other appropriate agencies determine that new needs for action, changed conditions or new alternatives having different environmental effects must be analyzed.  At that time, this analysis and document would be supplemented pursuant to NEPA.  Review of the EA would be conducted  each year at the time of the wildlife damage management work planning process by the Florida WS, NPS, USFWS, DOD, FDEP, FFWCC, and other appropriate agencies and/or entities to ensure that the EA is sufficient.

**Site Specificity.**  This EA addresses all lands under cooperative agreement, agreement for control, WS Work Plans or other comparable documents in Florida.  These lands are under the jurisdiction of federal, state, county, municipal and private administration/ownership. It also addresses the impacts of predator damage management on areas where additional agreements may be signed in the future. Because the proposed action is to reduce predator damage and because the program's goals and directives are to provide services when requested, within available funding and workforce, it is conceivable that additional wildlife damage management efforts could occur. Thus, this EA anticipates this potential expansion and analyzes the impacts of such efforts as part of the program. This EA emphasizes major issues as they relate to specific areas whenever possible, however, many issues apply whenever wildlife damage and resulting management occur, and are treated as such. The standard ADC Decision Model (Slate et al. 1992, USDA 1994) would be the site-specific procedure for individual actions conducted by WS in Florida.

**Summary of Public Involvement.** Issues related to the proposed action were initially developed by an interdisciplinary team process involving the USFWS, NPS, DOD, FDEP, and the FFWCC. A Multi-agency Team of WS, USFWS, NPS, DOD, FDEP, and FFWCC personnel refined these issues, prepared objectives and identified preliminary alternatives. Due to interest in the Florida WS Program, the Multi-agency Team concurred that Florida WS include an invitation for public comment in the initial development of this EA process. An invitation for public comment letter containing issues, objectives, preliminary alternatives, and a summary of the need for action was sent to 27 individuals or organizations for their input.

## 1.6     AUTHORITY AND COMPLIANCE

### 1.6.1     Authority of Federal Agencies in Wildlife Damage Management in Florida

#### Wildlife Services Legislative Mandate - Animal Damage Control Act of 1931

The primary statutory authority for the Wildlife Services program is the *Animal Damage Control Act of 1931*, which provides that:

*"The Secretary of Agriculture is authorized and directed to conduct such investigations, experiments, and tests as he may deem necessary in order to determine, demonstrate, and promulgate the best methods of eradication, suppression, or bringing under control on national forests and other areas of the public domain as well as on State, Territory or privately owned lands of mountain lions, wolves, coyotes, bobcats, prairie dogs, gophers, ground squirrels, jackrabbits, brown tree snakes and other animals injurious to agriculture, horticulture, forestry, animal husbandry, wild game animals , furbearing animals, and birds, and for the protection of stock and other domestic animals through the suppression of rabies and tularemia in predatory or other wild animals; and to conduct campaigns for the destruction or control of such animals. Provided that in carrying out the provisions of this Section, the Secretary of Agriculture may cooperate with States, individuals, and public and private agencies, organizations, and institutions."*

Since 1931, with the changes in societal values, WS policies and its programs place greater emphasis on the part of the Act discussing "bringing (damage) under control", rather than "eradication" and "suppression" of wildlife populations. In 1988, Congress strengthened the legislative mandate of WS with the Rural Development, Agriculture, and Related Agencies Appropriations Act. This Act states, in part:

*"That hereafter, the Secretary of Agriculture is authorized, except for urban rodent control, to conduct activities and to enter into agreements with States, local jurisdictions, individuals, and public and private agencies, organizations, and institutions in the control of nuisance mammals and birds and those mammals and birds species that are reservoirs for zoonotic diseases, and to deposit any money collected under any such agreement into the appropriation accounts that incur the costs to be available immediately and to remain available until expended for Animal Damage Control activities."*

## U.S. Department of Interior, Fish and Wildlife Service Legislative Mandate

The U. S. Fish and Wildlife Service's (USFWS) authority for action is based on the Migratory Bird Treaty Act of 1918 (as amended), which implements treaties with the United States, Great Britain (for Canada), the United Mexican States, Japan, and the Soviet Union. Section 3 of this Act authorized the Secretary of Agriculture:

"From time to time, having due regard to the zones of temperature and distribution, abundance, economic value, breeding habits, and times and lines of migratory flight of such birds, to determine when, to what extent, if at all, and by what means, it is compatible with the terms of the convention to allow hunting, taking, capture, killing, possession, sale, purchase, shipment, transportation, carriage, or export of any such bird, or any part, nest, or egg thereof, and to adopt suitable regulations permitting and governing the same, in accordance with such determinations, which regulations shall become effective when approved by the President".

The authority of the Secretary of Agriculture with respect to the Migratory Bird Treaty was transferred to the Secretary of the Interior in 1939 pursuant to Reorganization Plan No. II. Section 4(f), 4 Fed. Reg. 2731, 53 Stat. 1433.

> **CFR 50 Subchapter C - The National Wildlife Refuge System - Part 30 - Feral Animals - Subpart B-30.11 - Control of feral animals.** (a) Feral animals, including horses, burros, cattle, swine, sheep, goats, reindeer, dogs, and cats, without ownership that have reverted to the wild from a domestic state may be taken by authorized Federal or state personnel or by private persons operating under permit in accordance with applicable provisions of Federal or State law or regulation.

## U.S. Department of Interior, National Park Service Legislative Mandate.

The primary statutory authority for the National Park Service is provided in the *National Park Service Organic Act of 1916*. Through this act, Congress established the National Park Service and mandated that it "shall promote and regulate the use of the federal areas known as national parks, monuments, and reservations...by such means and measures as conform to the fundamental purpose of the said parks, monuments, and reservations, which purpose is to conserve the scenery and the natural and historic objects and the wild life therein and to provide for the enjoyment of the same in such manner and by such means as will leave them unimpaired for the enjoyment of future generations." The Organic Act authorizes the Secretary to promulgate rules and regulations necessary for the management of the parks. This authority, among others, provides the basis for the regulations in 36 CFR 1.

> **Endangered, Threatened, and Rare Species Management.** The NPS *Management Policies* prescribes management of endangered, threatened, and candidate species in

conformance with the *Endangered Species Act*, recovery plans, and other related documents. *Management Policies* states:

The National Park Service will identify and promote the conservation of all federally listed threatened, endangered, or candidate species within park boundaries and their critical habitats....The National Park Service also will identify all state and locally listed threatened, endangered, rare, declining, sensitive, or candidate species that are native to and present in the parks, and their critical habitats....All management actions for protection and perpetuation of special status species will be determined through the park's resource management plan. (4:11).

**Exotic Species Management.** NPS *Management Policies* addresses exotic species management mainly in the section on Exotic Plants and Animals (4: 11-12). In general, the NPS strives to protect and preserve all species of native flora and fauna within all management areas. Regarding exotic species, *Management Policies* states that:

Nonnative [exotic] plants and animals will not be introduced into natural zones except in rare cases where they are the nearest living relative of extirpated native species, where they are improved varieties of native species that cannot survive current environmental conditions, where they may be used to control established exotic species, or when directed by law or expressed legislative intent....

Management of populations of exotic plant and animal species, up to and including eradication, will be undertaken whenever such species threaten park resources or public health....High priority will be given to the management of exotic species that have a substantial impact on park resources and that can reasonably be expected to be successfully controlled. (4:12).

**U.S. Air Force - Policy Directive (AFPD) 32-70, Environmental Quality, and Department of Defense Instruction (DODI) 4715.3, Environmental Conservation Program.**

**Fish and Wildlife Management Component Plans (6.1.).** The fish and wildlife management component plan in the INRMP (Integrated Natural Resources Management Plan) addresses the management of game and nongame species on an installation......

Category I installations shall develop a fish and wildlife management component plan to the INRMP. To comply with the Sikes Act (16 USC 67 a-1[b]), United States military reservations must use professionally trained fish and wildlife management personnel to develop, implement, and enforce their fish and wildlife management programs (6.1.2.).

**Hunting, Fishing, and Trapping Programs (6.3.).** If practical, develop hunting, fishing, and trapping programs for recreation and wildlife population control.... The Sikes Act stipulates that these fees be used on the installation where they are collected, and must be used for the protection, conservation, and management of fish and wildlife, including habitat improvement and related activities.....

**Wildlife Damage Control (6.6.).** MAJCOMs (Major Commands) authorize emergency control measures only when wildlife endangers installation operations or the public health. The Animal and Plant Health Inspection Service (APHIS), the USFWS, and the state fish and wildlife agency should be notified as soon as practicable (6.6.2.).

**Regulatory Basis (7.1.).** The Endangered Species Act (Public Law 93-205) requires protection and conservation of federally listed T/E plants and animals and their habitats. Installations that know that they have T/E species or habitat critical for such species must include a T/E species component plan in the INRMP. An installation's overall ecosystem management strategy must provide for the protection and recovery of T/E species.

When practical, give the same protection to candidate species that you do for species that are already listed. Although the Endangered Species Act does not require it, give the same protection to state-listed T/E or rare species when practical (7.1.1.).

## 1.6.2    Compliance with Other Federal and State Statutes

Several federal laws, state laws, and state regulations regulate WS wildlife damage management. WS complies with these laws and regulations, and consults and cooperates with other agencies as appropriate.

**Florida Game and Freshwater Fish Commission (name was changed in 1999 to: Florida Fish and Wildlife Conservation Commission) - Authority to Manage State Wild Animal Life and Fresh Water Fish Life - Florida Constitution, Article IV, Section 9.**

"There shall be a game and fresh water fish commission, composed of five members appointed by the governor subject to confirmation by the senate for staggered terms of five years. The commission shall exercise the regulatory and executive powers of the state with respect to wild animal life and freshwater aquatic life, ......".

**Florida Department of Environmental Protection - Florida Park Service Authority**

> **Florida Statute - Chapter 258 - 258.037 - State Parks and Preserves - Part I - Policy of Division.** "It shall be the policy of the Division of Recreation and Parks: .... to acquire typical portions of the original domain of the state... and of such character as to emblemize the state's natural values; conserve these natural values for all time..."

> **Florida Administrative Code - Chapter 62D-2.014 & 62D-2.013 - Park Property and Resources & Hunting and Firearms.** "The Division may authorize the control of nuisance animals and may remove all exotic animals from parks by trapping and other necessary means for park resources management purposes. Such authorization shall be in the form of a license, permit, or contract negotiated by the parties or made pursuant to an advertised bid by the Division."

> **Resource Management Policy # 1 - Nuisance And Exotic Animals**

> I.    Nuisance Animals are individual animals of native species whose actions create special management problems. Examples of animal species from which nuisance cases may arise include raccoons, gray squirrels, poisonous snakes, and alligators...

>     A.    A potential threat to humans of physical injury (bites or scratches) or disease occurs due to abnormal or conditioned animal behavior patterns, including persistence in high public use areas.

B.  Unacceptable damage occurs to park facilities or other public or private property.

C.  Unacceptable damage occurs to valuable park natural resources, e.g., raccoons destroying sea turtle nests.

II.  The following management measures for resolving nuisance animal problems are listed in decreased order of preference....

D.  Humanely destroy nuisance animals.  Destruction of persistent nuisance animals should be dealt with on a case-by-case basis, requiring consultation with the Bureau of Natural and Cultural Resources, except in emergencies when immediate action must be taken to safeguard staff or visitors.  Parks that in previous seasons have experienced significant predation of sea turtle nests by raccoons, foxes, or coyotes should attempt to reduce those predator populations by relocation, if practicable, or humane destruction, if necessary, prior to the nesting season....

III.  Exotic animals are species not indigenous to Florida that occur here usually because of human-aided range expansion or translocation.  They include foreign species as well as free-ranging domesticated and feral animals.

IV.  Management measures to deal with exotic animals are as follows:

A.  Exotic animals shall be eliminated from parks by capture and removal, as is practicable, and if not, by humanely destroying individual animals.  Priority should be given to destructive and invasive species.  Relocation should occur to other properties only with an appropriate FGFWFC (FFWCC)  permit and the landowner's permission...

B.  Domestic animals owned as pets or livestock (e.g., dogs, cats, cattle):

2.  If no animal control facility exists within a reasonable distance.... and the animal poses a risk to park natural resources...the Park Manager may authorize the humane destruction of the animal in the park by park staff.

C.  Feral animals will be considered in the same manner as domestic animals... Feral hogs are covered under Standard Resource Management Procedure # 11, Feral Hog Removal.

**Standard Resource Management Procedures**

**Number 11 - Feral Hog Removal**

Procedures

3.  Hogs may be removed by trapping, catch dogs, or by shooting.  Trapping may be by any humane method.

7.  Agreements with Governmental Agencies or Private Nonprofit Organizations:  When appropriate, the District Manager may authorize hog removal by other governmental agencies...  To reduce or eliminate hogs from state park lands...

### Number 10 - Coyote Control

Procedures

> Coyotes are not protected on Department-managed lands... Control of coyotes is warranted in specific situations where they are known to be killing listed species.

## National Environmental Policy Act (NEPA).

Environmental documents pursuant to NEPA must be completed before work plans consistent with the NEPA decision can be implemented. WS also coordinates specific projects and programs with other agencies. The purpose of these contacts is to coordinate any wildlife damage management that may affect resources managed by these agencies or affect other areas of mutual concern.

## Endangered Species Act (ESA).

It is federal policy, under the ESA, that all federal agencies shall seek to conserve endangered and threatened species and shall utilize their authorities in furtherance of the purposes of the Act [Sec. 7(a)(1)]. WS conducts Section 7 consultations with the FWS to use the expertise of the FWS to ensure that "any action authorized, funded or carried out by such an agency. . . is not likely to jeopardize the continued existence of any endangered or threatened species. . . Each agency shall use the best scientific and commercial data available" [Sec. 7(a)(2)].

## Migratory Bird Treaty Act (MBTA).

The MBTA provides the USFWS regulatory authority to protect species of birds that migrate outside the United States. The law prohibits any "take" of the species, except as permitted by the USFWS or by federal agencies within the scope of their authority; therefore the USFWS issues permits for managing wildlife damage situations. Historically, the MBTA permit requirements did not apply to Federal agencies. However, based on recent advise received from the USDA Office of General Council, WS will receive a depredation permit before any control activities are conducted that involves the "take" of a species protected under the MBTA. Therefore, if WS conducts control activities involving the "take" of a species protected by the MBTA, a USFWS permit will be obtained prior to the implementation of any operational control activities on a MBTA protected species. Additionally, WS actions are consistent with what is allowed under 50 Code of Federal Regulations, Part 21, developed by the USFWS. WS may conduct control activities under the authority of USFWS permits issued to individuals or other federal and state agencies when listed as a named agent on the permits. Furthermore, if state agencies are to assist WS in taking migratory birds, then those state agencies are required by MBTA to obtain a permit.

## Federal Insecticide, Fungicide, and Rodenticide Act (FIFRA).

FIFRA requires the registration, classification, and regulation of all pesticides used in the United States. The United States Environmental Protection Agency (EPA) is responsible for implementing and enforcing FIFRA. All chemical methods integrated into the WS program in Florida are registered with and regulated by the EPA, FDA, and the Florida Department of Agriculture and Consumer Services [(FDACS) Chapter 487.155, Florida Statutes], and used by WS in compliance with labeling procedures and requirements.

20

## Investigational New Animal Drug (INAD).

The Food and Drug Administration (FDA) grants permission to use investigational new animal drugs [21 Code of Federal Regulations (CFR), Part 511]. Alpha chloralose is now classified as an animal drug (21 CFR 510) and cannot be purchased from any source except WS. The FDA authorization allows WS to use alpha chloralose to capture geese, ducks, coots, and pigeons. FDA acceptance of additional data will allow WS to consider requesting an expansion in the use of alpha chloralose to include other species.

## Environmental Justice and Executive Order 12898. Federal Actions to Address Environmental Justice in Minority Populations and Low - Income Populations.

Environmental Justice has been defined as the pursuit of equal justice and equal protection under the law for all environmental statutes and regulations without discrimination based on race, ethnicity, or socioeconomic status. Executive Order 12898 requires Federal agencies to make Environmental Justice part of their mission, and to identify and address disproportionately high and adverse human health and environmental effects of Federal programs, policies and activities on minority and low-income persons or populations. A critical goal of Executive Order 12898 is to improve the scientific basis for decision-making by conducting assessments that identify and prioritize environmental health risks and procedures for risk reduction. Environmental Justice is a priority both within the APHIS and WS. APHIS plans to implement Executive Order 12898 principally through its compliance with the provisions of NEPA.

WS activities are evaluated for their impact on the human environment and compliance with Executive Order 12898 to ensure Environmental Justice. WS personnel use wildlife damage management methods as selectively and environmentally conscientiously as possible. All chemicals used by APHIS-WS are regulated by the EPA through the Federal Insecticide, Fungicide and Rodenticide Act (FIFRA), FDA, FDACS, Memorandum Of Understanding (MOU) with Federal natural resource managing agencies, and by ADC Directives. Based on a thorough Risk Assessment, APHIS concluded that when WS program chemicals are used following label directions, they are highly selective to target individuals or populations, and such use has negligible impacts on the environment (USDA 1994, Appendix P). The WS operational program properly disposes of any excess solid or hazardous waste. It is not anticipated that the proposed action would result in any adverse or disproportionate environmental impacts to minority and low-income persons or populations.

## National Historic Preservation Act (NHPA) of 1966. As amended.

The National Historic Preservation Act (NHPA) of 1966, and its implementing regulations (36 CFR 800), requires federal agencies to: 1) determine whether activities they propose constitute "undertakings" that can result in changes in the character or use of historic properties and, 2) if so, to evaluate the effects of such undertakings on such historic resources and consult with the State Historic Preservation Office regarding the value and management of specific cultural, archaeological and historic resources, and 3) consult with appropriate American Indian Tribes to determine whether they have concerns for traditional cultural properties in areas of these federal undertakings. WS actions on tribal lands will be conducted only at the tribe's request and under signed agreement; thus, the tribes will have control over any potential conflict with cultural resources on tribal properties. WS activities, as described under the proposed action, do not cause ground disturbances nor do they otherwise have the potential to significantly affect visual, audible, or atmospheric elements of historic properties and are thus not undertakings as defined by the NHPA. Predator damage management could benefit historic properties if such properties were being damaged by feral hogs or other destructive predator species. In those cases, the officials responsible for management of such properties would make the request and would have decision-making authority over the methods to be used. WS has determined predator damage management actions are not undertakings as defined by the NHPA because such actions do not have the potential to result in changes in the character or use of historic properties. A copy of this EA will be provided to any American Indian tribe in the State that expresses a concern or interest in the proposed WS action and/or prior to any WS activity proposed to be conducted on reservation lands.

## 1.7    A PREVIEW OF THE REMAINING CHAPTERS IN THIS EA

This EA is composed of five chapters and two appendices. Chapter 2 discusses and analyzes the issues and affected environment. Chapter 3 contains a description of each alternative, alternatives not considered in detail, and mitigation and SOPs. Chapter 4 analyzes the environmental impacts associated with each alternative considered in detail. Chapter 5 contains the list of preparers of this EA. Appendix A is the literature cited in the EA and Appendix B is the glossary of the EA.

# CHAPTER 2:  ISSUES AND AFFECTED ENVIRONMENT

## INTRODUCTION

Chapter 2 contains a discussion of the issues, including those that will receive detailed environmental impacts analysis in Chapter 4 (Environmental Consequences), and those that were used to develop mitigation measures and SOPs, and the issues that will not be considered in detail with rationale. Pertinent portions of the affected environment will be included in this chapter in the discussion of issues used to develop mitigation measures. Additional affected environments will be incorporated into the discussion of the environmental impacts in Chapter 4.

Issues are concerns of the public and/or of professional communities about potential environmental problems that might occur from a proposed federal action. Such issues must be considered in the NEPA decision process. Issues relating to the management of wildlife damage were raised during the scoping process in preparing the programmatic WS FEIS (USDA 1994) and were considered in the preparation of this EA. These issues are fully evaluated within the FEIS, which analyzed specific data relevant to the Florida WS Program.

## 2.1    AFFECTED ENVIRONMENT

The areas of the proposed action include beach and dune coastal ecosystems along the Atlantic and Gulf Coasts of Florida and inland areas incurring significant hog damage. All areas proposed for current and future predator damage management are areas where the said T&E species are incurring damage by predators. Control areas may include federal, state, county, city, private, or other lands, where WS assistance has been requested by a landowner or manager to control predator damage to T&E species. The control areas would also include property in or adjacent to identified sites where predation activities could cause damage to T&E species at breeding/nesting sites. Predator damage control would be conducted when requested by a landowner or manager, and only on properties with a Cooperative Agreement with Wildlife Services.

## 2.2    ISSUES  ADDRESSED IN DETAIL IN CHAPTER 4

Following are issues that have been identified as areas of concern requiring consideration in this EA.

- Effects of Predation on Resources Protected, Including Native Wildlife and Plant Species
- Effects on Target Species Populations
- Effects of Control Methods on Nontarget Species Populations, Including T&E Species
- Humaneness of Control Methods
- Effects of Control Methods on Human Health and Safety
- Effects on the Aesthetic Values of Targeted Species and Protected T&E Species

Potential environmental impacts of the Proposed Action and Alternatives in relation to these issues are discussed in Chapter 4. As part of this process, and as required by the Council on Environmental Quality (CEQ) and APHIS-NEPA implementing regulations, this document and its Decision are being made available to the public through "Notices of Availability" (NOA) published in local media and through direct mailings of NOA to parties that have specifically requested to be notified. New issues or alternatives raised after publication of public notices will be fully considered to determine whether the EA and its Decision should be revisited and, if appropriate, revised. Following the evaluation and/or the

1

incorporation of any additional information received by WS into this EA , WS will release a Decision Notice and Finding Of No Significant Impact (FONSI) for this EA to the public.

### 2.2.1 Effects of Predation on Resources Protected, Including Native Wildlife and Plant Species

Some people are concerned about the damaging effects that native wildlife and feral animals are having on the recovery of State and Federally Endangered, Threatened, Species of Special Concern, and Candidates of Fauna and Flora within Florida. These protected resources are commonly referred to as "listed species". These people are concerned as to whether the proposed action or any of the alternatives would reduce such damage to acceptable levels.

### 2.2.2 Effects on Target Species Populations

Some persons are concerned that the proposed action or any of the alternatives would result in the loss of local raccoon, fox, coyote, feral hog, and armadillo populations or could have a cumulative adverse impact on regional or statewide populations. Furthermore, some persons are concerned that the proposed action or any of the alternatives would result in adverse impacts to feral/free-ranging cats and dogs.

### Florida Fish & Wildlife Conservation Commission - Furbearer Data

The Florida Fish and Wildlife Conservation Commission, Furbearer Biologist, was consulted in regards to any potential or suspected adverse impacts that would result from the WS's proposed action. It was determined that the WS's proposed action would not significantly impact any of the species proposed for damage management and that the affect would only be localized and would not adversely affect adjacent predator populations.

Harvest records of furbearing species in Florida was obtained from the FFWCC, for 1992-1998 (Table 2-1). From this information, it would appear that the trapping of furbears in Florida has been very limited over the last seven years, and the major factor driving fur trapping is the market price of Florida fur (Table 2-2). This trend is also apparent in the number of trappers that are registered to trap furbears in Florida (Table 2-3). In regard to the best information available, it would appear that furbears receive little pressure from trappers in Florida, and that all species being considered for predator management are abundant, if not numerous throughout the coastal regions of the state. As a result, it is not believed that the WS's proposed action will impact the target or nontarget species on a county, regional, or statewide level.

Table 2-1. Florida furbearer harvest summary for 1992-1998.

| YEAR | 1992-93 | 1993-94 | 1994-95 | 1995-96 | 1996-97 | 1997-98 |
|---|---|---|---|---|---|---|
| OTTER | 105 | 213 | 175 | 245 | 238 | 342 |
| BOBCAT | 45 | 41 | 50 | 51 | 27 | 34 |
| MINK | 3 | 0 | 1 | 1 | 1 | 0 |
| RACCOON | 1345 | 1503 | 2286 | 2606 | 3610 | 2712 |
| OPOSSUM | 0 | 3 | 40 | 40 | 66 | 4 |
| BEAVER | 0 | 0 | 4 | 4 | 11 | 53 |
| NUTRIA | 0 | 0 | 0 | 0 | 0 | 0 |
| SKUNK | 0 | 0 | 0 | 0 | 0 | 0 |
| COYOTE | 0 | 0 | 0 | 0 | 1 | 1 |

Table 2-2. Average fur prices ($$) paid for Florida pelts (based on a sub-sample of dealers).

| YEAR | 1992-93 | 1993-94 | 1994-95 | 1995-96 | 1996-97 | 1997-98 |
|---|---|---|---|---|---|---|
| OTTER | 25 | 30 | 25 | 35 | 30 | 25 |
| BOBCAT | 15 | 15 | 15 | 20 | 15 | 9 |
| RACCOON | 5 | 5 | 5 | 6 | 6 | 8 |
| OPOSSUM | 1.5 | 1 | 1 | 1 | 1.5 | 1 |
| BEAVER | - | 3 | 3 | 3 | 8 | 9 |
| MINK | - | 7 | 7 | 7 | 7 | - |

Table 2-3. Fur trapping licenses sold in Florida between 1992-1998.

| YEAR | 1992-93 | 1993-94 | 1994-95 | 1995-96 | 1996-97 | 1997-98 |
|---|---|---|---|---|---|---|
| NON-RESIDENT | 0 | 0 | 0 | 1 | 0 | 0 |
| RESIDENT | 227 | 225 | 232 | 228 | 216 | 288 |
| TOTAL | 227 | 225 | 232 | 229 | 216 | 288 |

## WS Predator Damage Management in Florida

Since 1996, WS has conducted predator management operations, in regards to the protection of T&E species, in four areas of the state. These areas consist of both state and federally managed lands and include St. Joseph State Park, Gulf Islands National Seashore, Eglin Air Force Base, and Hobe Sound National Wildlife Refuge. Over a 4-year period, seven coyotes, thirteen red foxes, forty-nine raccoons, and eight armadillos have been removed from these four areas (numbers were combined for the four sites); nontarget species take included two white-tailed deer. Four additional nontarget species were trapped and released unharmed (1 - alligator, 6 - raccoons, 1 - bobcat, 1 - dog). Total WS take for this period was 79 animals; less than 3 % were nontarget species. The total nontarget catch for the same period was 11 animals; more than 95 % of these animals were released unharmed. It is important to point out that the result of predator management at these sites is the significant reduction of predation incurred by T&E species using these areas.

Based on the best information available and the species proposed for control work, WS does not anticipate that its limited program will significantly effect any species, regional population, statewide population, or effect species populations in adjoining states (no significant cumulative impact). The species proposed for control are non-migratory and considered common to abundant; in many areas raccoon numbers are great enough to create a nuisance and health hazard. Based on trapping data, none of the species proposed for control are heavily impacted by trappers. When compared to other states, with the exception of habitat loss due to development, there is little to no impact to these species in Florida. It is possible that WS control operations may increase the health of target species' populations in the localized work areas.

### 2.2.3 Effects of Control Methods on Nontarget Species Populations, Including T&E Species

A common concern among members of the public and wildlife professionals, including WS personnel, is the potential for control methods used in the proposed action or any of the alternatives to inadvertently capture or remove nontarget animals or potentially cause adverse impacts to nontarget species populations, particularly T&E species. WS's mitigation and SOPs are designed to reduce the effects on nontarget species' populations and are presented in Chapter 3. To reduce the risks of adverse affects to nontarget species, WS would select damage management methods that are as target-selective as possible or apply such methods in ways to reduce the likelihood of capturing nontarget species. Before initiating trapping, WS would select trapping locations which are extensively used by the target species and use baits or lures which are preferred by the target species.

## WS Predator Damage Management in Florida

Since 1996, WS has conducted predator management operations, in regards to the protection of T&E species, in four areas of the state. These areas consist of both state and federally managed lands and include St. Joseph State Park, Gulf Islands National Seashore, Eglin Air Force Base, and Hobe Sound National Wildlife Refuge. Over a 4-year period, seven coyotes, thirteen red foxes, forty-nine raccoons, and eight armadillos have been removed from these four areas (numbers were combined for the four sites); nontarget species take included two white-tailed deer. Four additional nontarget species were trapped and released unharmed (1 - alligator, 6 - raccoons, 1 - bobcat, 1 - dog). Total WS take for this period was 79 animals; less than 3 % were nontarget species. The total nontarget catch for the same period was 11 animals; more than 95 % of these animals were released unharmed. It is important to point out that the result of predator management at these sites is the significant reduction of predation incurred by T&E species using these areas.

WS has determined that the proposed action has a low probability of adversely affect any species protected under the Florida Endangered Species Act and United States Endangered Species Act. This determination was concurred by WS biologists and other state and federal agencies involved in managing the said protected species.

### 2.2.4   Humaneness of Control Techniques

The issue of humaneness, as it relates to the killing or capturing of wildlife is an important, but very complex concept that can be interpreted in a variety of ways. Humaneness is a person's perception of harm or pain inflicted on an animal, and people may perceive the humaneness of an action differently. Animal welfare organizations are concerned that some methods used to manage wildlife damage expose animals to unnecessary pain and suffering. Research suggests that with some methods, such as restraint in leghold traps, changes in the blood chemistry of trapped animals indicate "stress." Blood measurements indicated similar changes in foxes that had been chased by dogs for about five minutes as those restrained in traps (USDA 1994). However, such research has not yet progressed to the development of objective, quantitative measurements of pain or stress for use in evaluating humaneness.

The decision making process involves tradeoffs between managing damage and the aspect of humaneness. The challenge in coping with this issue is how to achieve the least amount of animal suffering with the constraints imposed by current technology, yet provide sufficient damage management to resolve problems.

WS has improved the selectivity of management devices through research and development such as pan tension devices for traps and breakaway snares. Research is continuing to bring new findings and products into practical use. Until such time as new findings and products are found to be practical, a certain amount of alleged animal suffering will occur if management objectives are to be met in those situations where nonlethal control methods are not practical.

WS personnel in Florida are experienced and professional in their use of management methods. Consequently, control methods are implemented in the most humane manner possible under the constraints of current technology. Mitigation measures and SOPs used to maximize humaneness are listed in Chapter 3.

### 2.2.5   Effects of Control Methods on Human Health and Safety

A common concern is whether the proposed action or any of the alternatives pose an increased threat to human health and safety. Specifically, there is concern that the lethal methods of predator removal (i.e., shooting) may be hazardous to people.

Firearm use in wildlife damage control can be a publicly sensitive issue. Safety issues related to the misuse of firearms and the potential human hazards associated with firearms use are concerns both to the public and WS. To ensure safe use and awareness, WS employees who use firearms to conduct official duties are required to attend an approved firearms safety and use training program within 3 months of their appointment and a refresher course every 3 years afterwards (WS Directive 2.615). WS employees who use firearms as a condition of employment, are required to sign a form certifying that they meet the criteria as stated in the *Lautenberg Amendment* which prohibits firearm possession by anyone who has been convicted of a misdemeanor crime of domestic violence. Additionally, WS runs thorough background checks on all new employees entering the agency and the Florida WS Program conducts annual firearms training for its personnel.

### 2.2.6   Effects on the Aesthetic Values of Targeted Species and Protected T&E Species

The human attraction to animals has been well documented throughout history and started when humans began domesticating animals. The American public shares a similar bond with animals and/or wildlife in general, and today a large percentage of American households have pets. However, some people may consider individual wild animals and birds as "pets" or exhibit affection toward these animals, especially people who enjoy coming in contact with wildlife. Therefore, the public reaction is variable and mixed to wildlife damage management because there are numerous philosophical, aesthetic, and personal attitudes, values, and opinions about the best ways to manage conflicts/problems between humans and wildlife.

There is some concern that the proposed action or the alternatives would result in the loss of aesthetic benefits to the public, resource owners, or neighboring residents. Wildlife generally is regarded as providing economic, recreational, and aesthetic benefits (Decker and Goff 1987), and the mere knowledge that wildlife exists is a positive benefit to many people. Aesthetics is the philosophy dealing with the nature of beauty, or the appreciation of beauty. Therefore, aesthetics is truly subjective in nature, dependent on what an observer regards as beautiful.
Wildlife populations provide a wide range of social and economic benefits (Decker and Goff 1987). These include direct benefits related to consumptive and non-consumptive use (e.g., wildlife-related recreation, observation, harvest, sale, etc.), indirect benefits derived from vicarious wildlife related experiences (e.g., reading, television viewing, etc.), and the personal enjoyment of knowing wildlife exists and contributes to the stability of natural ecosystems [e.g.,

ecological, existence, bequest values (Bishop 1987)]. Direct benefits are derived from a user's personal relationship to animals and may take the form of direct consumptive use (using parts of or the entire animal) or non-consumptive use [viewing the animal in nature or in a zoo, photography (Decker and Goff 1987)]. Indirect benefits or indirect exercised values arise without the user being in direct contact with the animal and come from experiences such as looking at photographs and films of wildlife, reading about wildlife, or benefiting from activities or contributions of animals such as their use in research (Decker and Goff 1987). Indirect benefits come in two forms: bequest and pure existence (Decker and Goff 1987). Bequest is providing for future generations and pure existence is merely knowledge that the animals exist (Decker and Goff 1987).

Some people have an idealistic view of wildlife and believe that all wildlife should be captured and relocated to another area to alleviate damage or threats to protected resources. Some people directly affected by the problems caused by wildlife strongly support removal. Individuals not directly affected by the harm or damage may be supportive, neutral, or totally opposed to any removal of wildlife from specific locations or sites. Some people totally opposed to predator damage management want WS to teach tolerance for damage and threats caused by wildlife, and that wildlife should never be killed. Some of the people who oppose removal of wildlife do so because of human-affectionate bonds with individual wildlife. These human-affectionate bonds are similar to attitudes of a pet owner and result in aesthetic enjoyment.

Florida WS only conducts predator damage management at the request of the affected property owner or resource manager. If WS received requests from an individual or official for predator damage management, WS would address the issues/concerns and consideration would be given as to the extent of WS involvement. Management actions would be carried out in a caring, humane, and professional manner.

## 2.3    ISSUES USED TO DEVELOP MITIGATION

### 2.3.1    Environmental Justice and Executive Order 12898 - "Federal Actions to Address Environmental Justice in Minority Populations and Low-Income Populations".

Environmental Justice (EJ) is a movement promoting the fair treatment of all races, income, and culture with respect to the development, implementation, and enforcement of environmental laws, regulations, and policies. Fair treatment implies that no person or group of people should endure a disproportionate share of the negative environmental impacts resulting either directly or indirectly from the activities conducted to execute this country's domestic and foreign policies or programs. EJ has been defined as the pursuit of equal justice and equal protection under the law for all environmental statutes and regulations without discrimination based on race, ethnicity, or socioeconomic status. (The EJ movement is also known as Environmental Equity -- which is the equal treatment of all individuals, groups or communities regardless of race, ethnicity, or economic status, from environmental hazards).

Environmental Justice is a priority both within the USDA/APHIS and WS. Executive Order 12898 requires federal agencies to make EJ part of their mission, and to identify and address disproportionately high adverse human health and environmental effects of federal programs, policies, and activities on minority and low-income persons or populations. A critical goal of Executive Order 12898 is to improve the scientific basis for decision-making by conducting assessments that identify and prioritize environmental health risks and procedures for risk

6

reduction. APHIS-WS developed a strategy that: 1) identifies major programs and areas of emphasis to meet the intent of the Executive Order, 2) minimize any adverse effects on the human health and environment of minorities and low-income persons or populations, and 3) carries out the APHIS mission. To that end, APHIS operates according to the following principles: 1) promote outreach and partnerships with all stakeholders, 2) identify the impacts of APHIS activities on minority and low-income populations, 3) streamline government, 4) improve the day-to-day operations, and 5) foster nondiscrimination in APHIS programs. In addition, APHIS plans to implement Executive Order 12898 through its compliance with the provisions of NEPA.

All APHIS-WS activities are evaluated for their impact on the human environment and compliance with Executive Order 12898 to insure EJ. WS personnel use wildlife damage management methods as selectively and environmentally conscientiously as possible. All chemical used by APHIS-WS are regulated by the EPA through FIFRA, by the FDACS, by MOUs with federal natural resource management agencies, and program directives. Based on a thorough Risk Assessment, APHIS concluded that when WS program chemicals are used following label directions, they are selective to target individuals or populations and such use has negligible impacts on the environment (USDA 1994, Appendix P). The APHIS-WS operational program, discussed in this document, properly disposes of any excess solid or hazardous waste. It is not anticipated that the proposed action would result in any adverse or disproportionate environmental impacts to minority or low-income persons or populations.

### 2.3.2 Protection of Children from Environmental Health and Safety Risks (Executive Order 13045).

WS prioritizes the identification and assessment of environmental health and safety risks that may disproportionately affect children. Children may suffer disproportionately from environmental health and safety risks for many reasons, including their physical and mental status. WS has concluded that the proposed management program would not create an environmental health or safety risks to children because the program would only make use of legally available and approved damage management methods applied where such methods are highly unlikely to adversely affect children.

## 2.4 ISSUES CONSIDERED BUT NOT IN DETAIL WITH RATIONALE

### 2.4.1 Legal Constraints on Implementation of Control.

WS is required to follow and adhere to all federal and state regulations. The methods proposed for use in predator damage management are all permitted by federal and state laws, or the appropriate exemptions/permits will be obtained.

### 2.4.2 Cost Effectiveness of Control Methods.

The methods determined to be most effective in controlling predator damage and proven to be most cost effective will receive the greatest application. Additionally, control operations may be constrained by cooperator monies and/or objectives and needs.

# CHAPTER 3 : ALTERNATIVES

## INTRODUCTION

Alternatives were developed for consideration using the ADC Decision Model as described in Chapter 2 (pages 20-35), Appendix J (Methods of Control), Appendix N (Examples of ADC Decision Model), and Appendix P (Risk Assessment of Wildlife Damage Control Methods Used by the USDA, Wildlife Services Program) of the *Animal Damage Control Program Final Environmental Impact Statement* (USDA 1994).

Chapter 3 contains a discussion of the project alternatives, including those that will receive detailed environmental impacts analysis in Chapter 4 (Environmental Consequences), and alternatives considered but not analyzed in detail, with rationale, and mitigation measures and SOPs for wildlife damage management techniques (WDM). Pertinent portions of the affected environment will be included in this chapter in the discussion of issues used to develop mitigation measures. Evaluation of the affected environments will be addressed in more detail in Chapter 4.

## ALTERNATIVES ANALYZED IN DETAIL

**Alternative 1 - No Action** - This alternative precludes any and all WDM activities by WS to protect T&E species in Florida. A natural resource manager or any other entity directed at preventing or reducing predation of sea turtle nests, crocodile nests, beach mice, and shorebirds could conduct WDM activities in the absence of WS involvement.

**Alternative 2 - Nonlethal Control Before Lethal Control** - This alternative would not allow the use or recommendation of lethal control by WS until all available nonlethal methods had been applied and determined to be inadequate in each damage situation.

**Alternative 3 - Nonlethal Control Only** - This alternative would involve the use and recommendation of nonlethal management techniques only by WS.

**Alternative 4 - Lethal Control Only** - This alternative would involve the use and recommendation of lethal management techniques only by WS.

**Alternative 5 - Integrated Wildlife Damage Management (the Proposed Action)** - This alternative would incorporate an integrated approach to wildlife damage management using components of the wildlife damage management techniques and methods addressed in Alternatives 2, 3, and 4, as deemed appropriate by WS and other participating entities.

## 3.1 DESCRIPTION OF THE ALTERNATIVES

### 3.1.1 Alternative 1 - No Action

This alternative precludes any and all WDM activities by WS to protect T&E species in Florida. A natural resource manager or any other entity directed at preventing or reducing predation of sea turtle nests, crocodile nests, beach mice, and shorebirds could conduct WDM practices in the absence of WS involvement.

1

### 3.1.2    Alternative 2 - Nonlethal Control Before Lethal Control

This Alternative would require that all methods or techniques described on 3.1.3 be applied and determined to be inadequate in each damage situation prior to the implementation of any of the methods or techniques described in 3.1.4. This would be the case regardless of the severity or intensity of predation on the resources proposed for protection in this EA.

### 3.1.3    Alternative 3 - Nonlethal Control Only

Exclusion devises and live trap and relocation of feral/free-ranging cats and dogs to local animal shelters are the only nonlethal control methods currently available for use to protect affected resources in Florida. Live trapping and relocation of other animal species would not be carried out by WS.

Nonlethal frightening devises have been determined to be unacceptable for use in any of the Alternatives. Frightening devises involving the use of electronic guards, pyrotechnics, propane cannons, and lights could potentially be used for temporary relief of predation; however, predators often become acclimated to such methods fairly rapidly and the use of these devices have the potential of adversely affecting the species needing protection. A detailed description of why frightening devices are not being considered in detail in this EA is found in Section 3.2.2.

Management strategies involving nonlethal methods would be limited to exclusion of sea turtle and crocodile nests by use of wire cages and the live trapping of feral/free-ranging cats and dogs.

#### Exclusion

Exclusion devices are applicable for use on sea turtle and crocodile nests only. They are not feasible nor effective for protecting nesting and wintering shorebirds, or any of the other species proposed for protection in this EA. This alternative would be used to deter predators from digging up individual sea turtle and crocodile nests. Excluders constructed of net wire fencing material, or comparable material, would be placed over the nests. The exclusion device currently in use consists of a 3 ½ foot square panel of net wire (2" by 4" mesh) securely anchored over each sea turtle nest when the nest is first laid, and once a nest has been located. When hatching is expected, the flat screen is sometimes replaced with a cage that protects hatchlings from predators. This cage restrains the hatchlings and personnel must release them.
Recommendations for modifying exclusion devices to increase their efficiency would be developed, as appropriate, for consideration.
Exclusion of crocodile nests using wire cages would follow a similar design as that of sea turtle nests; however, crocodile nests are often much more difficult to find than sea turtle nests and it is unlikely that excluders could be installed before a predator found the nest.

Excluding devices could be considered for protecting nesting birds, but it is feared that placing some sort of excluder over a nest would cause the parent birds to abandon the nest.

If any of the above exclusion devices are to be employed, it would be the responsibility of the natural resource manager to do so.

## Live Trapping/Relocation of Feral/Free-ranging Cats and Dogs

Live trapping and relocation of feral/free-ranging cats and dogs could be accomplished by the use of walk-in cage traps, leghold traps, or snares. These control devices are described in detail in Section 3.1.4. Cats and dogs would be relocated to the nearest animal shelter facility and would not, under any circumstance, be released back into the wild by Wildlife Services personnel.

### 3.1.4    Alternative 4 - Lethal Control Only

This alternative would allow the lethal removal of damage causing predators, including raccoons, foxes, coyotes, feral hogs, rats, and armadillos, involved in T&E species damage or predation, and those posing a predation threat to T&E species. Lethal control methods would be applied in all areas of control operations. Feral/free-ranging domestic cats and dogs that were captured in restraining devices would be taken to the nearest animal shelter. Predators (excluding free-ranging cats and dogs) would be euthanized on site in a humane manner utilizing AMVA approved methods and WS SOP's. Euthanization would occur by either injection with a WS approved drug or by shooting. Deceased animals would be buried or taken to a landfill, in accordance with WS policy and State Regulations. Unharmed and uninjured nontarget animals that could be safely handled, would be released on site.

Lethal methods of wildlife control are often very effective when used properly. Specific problem animals can be targeted and removed without negatively affecting the local population of a species (Bailey 1984). All control measures would be implemented in accordance with applicable Federal and State laws, and WS policy. Weather and environmental conditions permitting, all field equipment would be checked at least once each day. If daily checking is not possible, all control equipment would be removed from the site. Local population reduction of predators to reduce immediate predation losses and potential predation threats would be implemented by WS personnel with assistance from the participating natural resource managers. Target individuals would be lethally removed using the methods and techniques listed below.

> a.    Ground Shooting - This method would be used to selectively remove predators and feral hogs. Most shooting would be done in conjunction with night spotlighting or predator calling utilizing shotguns or rifles. Opportunistic shooting of target predators would occur in areas away from public use areas or during times when the public would not be present. This alternative would only be used in areas and at times which are deemed safe.

> b.    Leghold Traps - This method would be used to capture and restrain target predator species. Leghold traps, of the appropriate size and type, would be utilized to capture specific target animals. Leghold traps are a versatile and widely used control method. Placement of these traps is contingent upon the habits of the respective target species, habitat conditions, and presence of nontarget animals. Traps would be set in areas of high predator activity, including but not limited to pathways and watering holes. Traps could be placed as "baited" or "scented" sets, using an attractant consisting of fetid food, urine, or musk to attract the target animal to the trap location.

> Opposition to the use of leghold traps has increased in recent years due to public concern that the leghold trap inflicts unacceptable injuries to trapped animals. Research on the

No. 3 Victor Soft Catch leghold trap has demonstrated that coyotes can be successfully captured while producing only minor leg injuries (Phillips et al. 1996). Recent research comparing leg injuries associated with standard and modified Soft Catch leghold traps indicates that the addition of a "taos lightning" spring kit can further reduce injuries to captured animals and increase capture efficiency (Gruver et al. 1996). Soft Catch leghold traps modified with "taos lightening" springs kits may be used in some situations. Additionally, padded-jawed leghold traps may also be used to capture and restrain target species, however, WS will not limit trapping efforts to these devices.

c.    Walk-in Cage Traps - This method would be used to capture raccoons, armadillos, feral and free-ranging domestic cats and dogs, feral hogs, and in some instances, foxes. These traps would be set in areas where leghold traps could not be used, or when it was deemed more efficient to use them. Placement of walk-in cage traps is contingent upon the habits of the respective target species, habitat conditions, and presence of nontarget animals. Traps placed in travel lanes of the target animal, using location rather than attractants, are known as "blind sets". The "blind set" would be modified with two long boards placed on either side of the entrance of the trap to act as a funnel for trapping armadillos. More frequently, traps are placed as "baited" or "scented" sets, using an attractant consisting of fetid food, urine, or musk to attract the animal into the trap. Most feral/free-ranging cats would be trapped using these devices.

d.    Snares - Snares are capture devices comprised of a cable loop and a locking device. Most snares are equipped with a swivel to minimize cable twisting and breakage. Snares can be set as either lethal or live-capture devices. Neck snares are usually set as lethal devices. As a lethal device, neck snares are designed to tighten around an animal's neck as it passes through the device. Leg snares are live-capture devices meant to restrain the animal by tightening around the leg. Snares would be used as lethal and live-capture devices in narrow passageways and along well used predator pathways. Lethal snares would not be set to catch cats; however, live-capture snares may be used. Neck snares used in association with this project would incorporate break away locks.

e.    Denning - Denning is the practice of seeking out the dens of depredating coyotes, and foxes and eliminating the young, adults, or both to stop ongoing predation or prevent further predation. Denning would be used when appropriate and in specific cases where it has been determined necessary for alleviating a specific threat to sea turtle nests, crocodile nests, beach mice, and/or shorebirds.

The usefulness of denning, as a wildlife damage management method, is well known (Till and Knowlton 1983). However, it's use is limited because coyote and fox dens are difficult to locate and den use is restricted to approximately 2 to 3 months during the spring. Coyote and fox predation of available prey often increases during the spring and early summer because of the increased food requirements caused by the need to feed their pups. The removal of predator pups will often stop predation even when the adults are not taken. When the adults are taken and the den site is known, the pups are excavated and euthanized to prevent their starvation.

Denning activities would be confined to the natural resource managers area. Den hunting for adult coyotes, foxes and their young would be combined with calling and shooting as needed. Denning is highly selective for the target species and family groups responsible for damage.

4

### 3.1.5 Alternative 5 - Integrated Wildlife Damage Management (Proposed Action)

This alternative, the proposed action, would incorporate an integrated damage management program utilizing techniques and methods described in Alternatives 2, 3, and 4 to reduce sea turtle, crocodile, and shorebird nest predation by raccoons, foxes, coyotes, feral hogs, feral/free-ranging domestic dogs, and armadillos; predation threats to beach mice and adult shorebirds; predation threats sea turtle, crocodile, and shorebird hatchlings by raccoons, foxes, coyotes, feral hogs, and feral/free-ranging domestic cats and dogs. The integrated damage management program would also be effective in reducing the impacts of feral hogs on protected plants and animals. This strategy would incorporate the nonlethal and lethal control measures described in 3.1.3 and 3.1.4.

## 3.2   ALTERNATIVES CONSIDERED BUT NOT ANALYZED IN DETAIL WITH RATIONALE

### 3.2.1 Aversive Conditioning  (taste aversion) Alternative

The objective of aversive conditioning would be to feed egg predators a prey-like bait (eggs) laced with an aversive agent that causes them to become ill, resulting in the subsequent avoidance of the prey (eggs).

The use of any taste aversive agent would be experimental. No compounds are currently registered by the Environmental Protection Agency (EPA) for use in this situation. While some aversive conditioning studies involving raccoons and ravens have proven successful, results with coyotes, wild hogs, and armadillos have been less conclusive. To be successful the predator must be enticed to eat the egg baits; the predator aversive agent used must induce enough discomfort to condition the predator to avoid the baits; and this avoidance must be transferred to sea turtle and shorebird nests. Furthermore, the avoidance must persist long enough without reinforcement for this method to offer realistic protection to sea turtle, crocodile, and shorebird eggs. This method would not address the problem with predation on beach mice, shorebirds, nor sea turtle and crocodile hatchlings.

### 3.2.2   Frightening Devices Alternative

Frightening devices such as electronic guards, pyrotechnics, propane cannons, and lights can be used to temporarily alleviate predation. The effectiveness of these devices depends upon the individual predator's fear of, and subsequent aversion to the offensive stimuli. Once a predator habituates to these stimuli, it often resumes its normal activities and movements.

The continuous and prolonged utilization of artificial lighting along the beach could have a negative impact on sea turtle, crocodile, and shorebird nesting activity, and endangered beach mice foraging. The use of artificial lighting may deter female sea turtles (Witherington and Martin 1996) and shorebirds, discouraging them from nesting at historic nesting sites. In addition, newly hatched sea turtles are strongly attracted to light sources (Raymond 1984, Witherington 1995, Witherington 1991). This disorientation could lead to increased mortality due to predation, dehydration, and exhaustion. Lights could inhibit the foraging behavior of beach mice, since they forage during nighttime hours.

The impact of noise resulting from the use of electronic guards, pyrotechnics, and propane exploders in sea turtle and crocodile nesting areas is unknown. There are indications that the noise and harassment associated with increasing boat and jet ski traffic may stress sea turtles that are feeding, mating, or waiting to nest near popular beaches. Noise associated with the above devices, potentially could impact all animal species proposed for protection in this EA.

After consultation with the FPS and the USFWS, it was decided that this method was unacceptable for use during the sea turtle nesting season (May 1 to October 31), because of the potential impacts to adult nesting and hatchling sea turtles. This method could be used outside of the turtle nesting season from November 1 to April 30; however, the foraging activities of the beach mouse and wintering shorebirds would still be effected by the lights and noise from the frightening devices during this period. Also, using frightening devices during this time would not prevent predation of sea turtle and shorebird nests during nesting season.

Due to the public nature of the Florida coastal environs, and the presence of overnight campers, the use of electronic guards, pyrotechnics, and propane exploders would negatively impact the serene environment. The exclusive use of frightening devices in a manner compatible with park management and sea turtle nesting requirements would not reduce predation to an acceptable level.

### 3.2.3 Population Reduction (trap/translocate) Alternative

This alternative would allow the live capture of raccoons, foxes, coyotes, feral /free-ranging domestic cats and dogs, feral hogs, and armadillos using cage traps, leg snares, and/or leghold traps. Captured predators would be tranquilized and translocated to other areas.

The FWC, Title 39-4.005 (*Introduction of Foreign Wildlife or Freshwater Fish or Carriers of Disease*) does not allow the transportation of non-indigenous wildlife into or within the State of Florida. For the scope of this EA, this includes feral hogs, cats, dogs, and coyotes. Additionally, relocation of live furbearers (i.e., raccoons, coyotes, foxes, opossums, skunks, nutria, beaver) or nonprotected wildlife (i.e., armadillos) is not permitted in Florida without a permit issued by the FFWCC (FWC, Title 39-24.002 and 39-6.002).

Relocation of wildlife is often viewed as inhumane and biologically unsound management, especially when the wildlife species being relocated is already abundant or common in an area. Relocated animals are forced into a new environment where they often have to compete for space and resources with already well established animals of the same species. Consequently, WS will not request a permit from the state in regards to relocating any of the species proposed for control work in this EA. If certain segments of the public demand relocation, then it will be up to that group(s) to acquire a permit from the state and relocate the animals (as outlined in the relocation permit).

### 3.2.4 Eradication and Long Term Population Suppression of Native Wildlife Alternative

Eradication and long term population suppression of native wildlife is not an objective or option considered by the Wildlife Services Program in Florida. Eradication of native wildlife populations or species is considered ecologically unsound by the Wildlife Services Program, and is not and will not be conducted by WS. Within the scope of this EA, it is the objective of WS to reduce predator numbers within local populations that are directly impacting state and/or federally listed species. However, this reduction will be restricted to problem animals, species, or populations, and will only be conducted with non-native problem species and non-listed native

6

carnivores/omnivores that have been identified as significant predators of listed species in this EA. Additionally, non-native species (i.e., feral hogs) that directly impact the habitats of the listed species will be managed to reduce habitat degradation in these areas and to reduce their impact on other sensitive native fauna, flora, and ecosystems.

### 3.2.5 Biological Control Alternative

Biological control is most commonly used to control select evasive plant and insect species. Very little effort has been devoted to the biological control of wildlife species listed in this EA for two reasons: 1) many of these species are native to the North American continent and biological control measures directed towards a wide spread species potential could have disastrous, uncontrollable effects on a species throughout its range and 2) any biological control measure directed towards a non-native or feral species could adversely affect some groups of animals presently in use for agriculture purposes, ranching, pets, etc. that are closely related to the target species.

## 3.3    MITIGATION AND SOP's FOR WILDLIFE DAMAGE MANAGEMENT TECHNIQUES

### 3.3.1    Mitigation Measures

Mitigation measures are any features of an action that serve to prevent, reduce, or compensate for impacts that otherwise might result from that action. The current WS Program, nationwide and in Florida, uses many such mitigation measures and these are discussed in detail in Chapter 5 of the FEIS (USDA 1994).

Some key mitigating measures pertinent to the proposed action and alternatives that are incorporated into WS's SOPs include the following.

The WS Decision Model, which is designed to identify effective wildlife damage management strategies and their impacts, is consistently used.

♦   Nontarget animals captured in leghold traps or snares are released unless it is determined by a WS Specialist that the animal will not survive and/or that the animal can not be released safely.

♦   Conspicuous, bilingual warning signs alerting people to the presence of traps and snares may be placed at major access points to areas where WS is conducting active predator management operations, if it has been determined that the presence of the signs would not impact the efficacy of the management program in an area.

♦   Reasonable and prudent alternatives and measures are established through consultation with the USFWS and implemented to avoid adverse impacts to T&E species.

♦   EPA-approved label directions are followed for all pesticide use. Currently, none are planned for use in the scope of this EA.

### 3.3.2    Additional Mitigation Measures and SOPs for Wildlife Damage Management Techniques

Some additional mitigating factors specific to the current program include the following:

7

- All WS Specialists who use restricted-use chemicals are trained and certified by WS personnel or others who are experts in the safe and effective use of these substances or are supervised by such qualified persons.

- Management actions are directed toward individuals, species, or localized populations, responsible for damage to the T&E species listed in this EA. Generalized or blanket suppression of predator populations across Florida will not be conducted.

- Although hazards to the public from control devices and activities are low according to a formal risk assessment (USDA 1994, Appendix P), hazards to the public and their pets are even further reduced by the fact that control activities are primarily conducted during nighttime hours and by trained wildlife damage management specialists.

## 3.4 ADDITIONAL MITIGATION MEASURES SPECIFIC TO THE ISSUES.

The following is a summary of additional mitigation measures that are specific to the issues listed in Chapter 2 of this document.

### Effects on Target Species Populations

- WS activities conducted to resolve predation damage in respect to T&E species are directed towards individual problem animals, or local populations or groups, and not towards the eradication of a species or population within an entire area, region, or ecosystem.

- WS lethal take (kill) data are regularly monitored by WS biologists and are compliant with the recommended or authorized levels of harvest allowed by the State of Florida (See Chapter 4).

### Effects on Nontarget Species

- WS activities conducted to resolve predation damage are directed towards individual problem animals, or local populations or groups. Any nontarget animals captured in snares, cage traps, or leghold traps will be released whenever it is possible.

- When conducting removal operations via shooting, WS will shoot only target species or animals and will not shoot an animal that can not be accurately identified.

- WS specialists use lures, trap placements (sets), and capture devices that are strategically placed at locations likely to capture a target animal and minimize the potential of nontarget animal captures.

### Effects on Human Health and Safety

- WS control operations will be conducted professionally and in the safest manner possible. Most trapping will be conducted away from areas of high human activity and when determined necessary, signs will be placed to warn the public of any potential hazards.

- WS predator management via shooting will be conducted professionally and in the safest manner possible. Shooting will be conducted during time periods when public activity and access to the control areas are restricted. WS personnel involved in shooting operations will be fully trained in the proper and safe application of this method.

**Humaneness of Methods Used by WS**

- WS specialists will be well trained in the latest and most humane devices/methods for removing problem wildlife.

- WS personnel attempt to dispatch captured target animals, slated for lethal removal, as quickly and humanely as possible. In most field situations, a precise shot to the brain using a small caliber firearm is performed. This method causes rapid unconsciousness followed by the cessation of heart and respirator functions, resulting in a humane and rapid death. This method is in concert with the American Veterinary Medical Association's (AVMA) definition of euthanasia.

- The WS's National Wildlife Research Centers (NWRC) are continually conducting research, with the goal, to improve the selectivity and humaneness of wildlife damage management devices used by WS personnel in the field.

## CHAPTER 4: ENVIRONMENTAL CONSEQUENCES

### INTRODUCTION

Chapter 4 provides information needed for making informed decisions on the predator damage management objectives outlined in Chapter 1 and the issues and affected environment discussed in Chapter 2. This chapter analyzes the environmental consequences of each alternative in relation to the issues identified for detailed analysis. This chapter analyzes the environmental consequences of each alternative in comparison with the No Action Alternative to determine if the real or potential impacts would be greater, lesser, or the same. Therefore, the No Action Alternative serves as the baseline for the analysis and the comparison of expected impacts among the alternatives. The analysis also takes into consideration WS mandates, directives, and the procedures used in the WS decision process (USDA 1994).

The following resource values within the State of Florida are not expected to be significantly impacted by any of the alternatives analyzed: soils, geology, minerals, water quality/quantity, flood plains, wetlands, critical habitats (areas listed in T&E species recovery plans), visual resources, air quality, prime and unique farmlands, aquatic resources, timber, and range. These resources will not be analyzed further.

### 4.1    Detailed Analysis of Environmental Impacts of the Alternatives

#### 4.1.1    Effects of Predation on Resources Protected, Including Native Wildlife and Plant Species

**Alternative 1. No Action**

Under this alternative, WS would not be involved in Wildlife Damage Management (WDM) to reduce predation to State and Federally listed species. Many species of listed wildlife would continue to incur potentially disastrous levels of predation from the predators proposed for management, provided that natural resource managers did not implement their own WDM program. Efforts to reduce or prevent predation by natural resource mangers or others could increase. This increase, potentially could result in impacts on the protected species populations to an unknown degree. Impacts on protected species under this alternative could be the same, less than, or more than those of the proposed action depending on the level of effort expended by the natural resource managers.

The No Action Alternative could lead to the continued predation of sea turtles, crocodiles, colonial nesting seabirds, and other listed species. Feral hog damage to rare and sensitive plants could continue at current levels, and potentially contribute to the extirpation of many of these species or populations. Long term and irreversible negative biological impacts could result to the species addressed in this EA.

**Alternative 2. Nonlethal Control before Lethal Control**

Under this alternative, WS would implement and recommend nonlethal control prior to the use of lethal methods. It is likely that many species of listed wildlife would continue to incur potentially high levels of predation from the predators proposed for management. It is probable,

1

in many situations, that by the time all nonlethal methods were attempted and determined to be ineffective, the protected resource could be heavily impacted by predation. Currently, the only nonlethal method recommended by WS is exclusion (i.e., wire mesh cages, electric fences, etc.). Mammalian species could not be protected through exclusionary devices and other nonlethal methods would not adequately reduce predation.

Feral hogs are considered the major wildlife species contributing to the decline of several rare plant species. This species is considered highly intelligent and capable of avoiding human interactions rather easily. With any type of human harassment, feral hogs become more wary of humans and exceedingly difficult to control. Often, hogs become nocturnal in areas with frequent human encounters. Consequently, the use of nonlethal techniques would make control efforts less effective and prolong damage to these plants. This alternative would likely be more effective at preventing or reducing depredation to listed species than Alternatives 1 and 3, but not as effective as Alternatives 4 and 5.

**Alternative 3. Nonlethal Control Only**

Under this alternative, WS would only implement and recommend nonlethal control methods. Nonlethal methods have proven (in many cases) to be ineffective at reducing predation to T&E species. This alternative would do nothing to protect the endangered beach mice, woodrats, cotton mice, marsh rabbits, and colonial nesting seabirds; therefore, predation would continue at the same intensity for all species proposed for protection in this EA.

The use of exclusion to deny predators access to sea turtle nests can reduce some predation losses. Most natural resource managers began utilizing exclusion devices in 1993. In past years, the wire exclusion devices have afforded adequate nest protection and most do not impede the movement of hatchling sea turtles from the nest site. As predator populations increased, it was noted that predators began to dig under the exclusion devices to get to the eggs. Recent studies have documented predator adaptation to these exclusion devices. These findings are causing concern to natural resource managers because predation rates are increasing as this newly learned behavior is passed on to progeny. Another problem associated with exclusion is the cost and effort expended to patrol the beach along sea turtle nesting sites, locate, and install exclusion devices for sea turtle nest protection. To further complicate matters, predators often find and destroy the nests before patrol personnel are able to locate them. Considering the current human resources available to the natural resource managers, it is not possible to reduce predation losses to an acceptable level by exclusion only.

Exclusion could potentially alleviate some predation to American crocodile nests; however, the logistics and expense of locating crocodile nests before depredating raccoons would be considerably difficult and impractical.

Feral hogs are considered the major wildlife species contributing to the decline of several rare native plant species. This species is considered highly intelligent and capable of avoiding human interactions rather easily. With any type of human harassment, feral hogs become more wary of humans and exceedingly difficult to control. Often, hogs become nocturnal in areas with frequent human encounters. Consequently, the use of nonlethal techniques would make control efforts less effective and prolong damage to these plants. Feral hog exclusion from large areas

2

and systems would be highly impractical, if not impossible. Exclusionary devices (i.e., electric fencing, large mesh fencing, etc.) could be implemented on very small areas with moderate success in protecting some populations of plants; however, this method would do nothing to protect rare animal species from hog predation.

This alternative potentially would be more effective at preventing or reducing predation to the listed species than Alternative 1, providing that some effective level of nonlethal management could be implemented. Otherwise, the effects on listed species from this alternative would be similar to Alternative 1. This alternative would not be as effective in reducing predation to listed species as Alternatives 2, 4 and 5.

**Alternative 4. Lethal Control Only**

Under this alternative, WS would implement and recommend lethal control methods without applying or considering nonlethal methods. In most situations, lethal methods would be applied as a result of unsuccessful attempts by natural resource managers to alleviate predator damage through nonlethal methods. Predation of protected resources would likely be reduced or eliminated under this alternative, providing that lethal control methods could be safely and effectively implemented. In situations where lethal control could not be conducted, because of safety concerns or local ordinances, predation rates could be expected to remain the same or increase. This alternative would likely be more effective at preventing or reducing predation to listed species than Alternatives 1, 2 and 3, if some effective level of lethal management could be implemented. Otherwise, effects on listed species from this alternative would be similar to Alternative 1. This alternative would likely not be as effective in reducing predation to listed species as Alternative 5.

**Alternative 5. Integrated Wildlife Damage Management (Proposed Action)**

Under this alternative, WS would incorporate select components from Alternatives 2, 3, and 4 into its WDM program. This alternative has the greatest potential of reducing predation to listed species because all potential nonlethal and lethal control alternatives and methods would be available for use and recommendation by WS.

**4.1.2   Effects on Target Species Populations**

**Alternative 1. No Action**

Under this alternative, WS would not be involved in Wildlife Damage Management (WDM) to reduce predation to State and Federally listed species. No impact would be experienced by any target species or population as a result of WS operations. However, predator impacts on T&E species would continue at the current rate throughout Florida, providing that natural resource managers did not implement their own WDM program. The No Action Alternative could negatively impact all species proposed for protection in this EA. Efforts by natural resource mangers and other entities to reduce or prevent depredations could increase, potentially resulting in impacts on target species populations to an unknown degree. Impacts on target species under this alternative could be the same, less than, or more than those of the proposed action depending on the level of effort expended by the natural resource managers.

**Alternative 2.  Nonlethal Control before Lethal Control**

Under this alternative, WS would implement nonlethal control prior to the use of lethal methods. As stated in Section 2.2.2, it is not likely that WS would negatively impact target species populations on a local, regional, or statewide scale under this alternative. Some local reduction in predator populations may occur in localized areas were lethal control activities are implemented, but not to an extent that predator species would be permanently extirpated from an area. Local and regional immigration and emigration of predator species would be expected to replace removed target animals after a relatively short period of time. Captured feral cats and dogs would be transported to the nearest animal shelter. Impacts under this alternative would be similar to Alternatives 4 and 5, providing that lethal control is implemented. Otherwise, impacts would be similar to Alternatives 1 and 3.

**Alternative 3.  Nonlethal Control Only**

Under this alternative, WS would only implement nonlethal control methods. WS would not directly impact target wildlife species under this alternative. Captured feral cats and dogs would be transported to the nearest animal shelter.

**Alternative 4.  Lethal Control Only**

Under this alternative, WS would implement and recommend lethal control methods without applying or considering nonlethal methods. In most situations, lethal methods would be applied as a result of unsuccessful attempts by natural resource managers to alleviate predator damage through nonlethal methods. As stated in Section 2.2.2, it is unlikely that WS would negatively impact target species populations on a local, regional, or statewide scale under this alternative. Some local reduction in predator populations may occur in localized areas were lethal control activities are implemented, but not to an extent that predator species would be permanently extirpated from an area. Local and regional immigration and emigration of predator species would likely replace removed target animals after a relatively short period of time. Captured feral cats and dogs would be transported to the nearest animal shelter. Impacts under this alternative would be similar to Alternatives 2 and 5.

**Alternative 5.  Integrated Wildlife Damage Management (Proposed Action)**

Under this alternative, WS would incorporate select components from Alternatives 2, 3, and 4 into its WDM program. As stated in Section 2.2.2, it is unlikely that WS would negatively impact target species populations on a local, regional, or statewide scale under this alternative. Some local reduction in predator populations may occur in localized areas were lethal control activities are implemented, but not to an extent that predator species would be permanently extirpated from an area. Local and regional immigration and emigration of predator species would be expected to replace removed target animals after a relatively short period of time. Captured feral cats and dogs would be transported to the nearest animal shelter. Impacts under this alternative would be similar to Alternatives 2 and 4.

### 4.1.3 Effects of Control Methods on Nontarget Species Populations, Including T&E Species

**Alternative 1. No Action**

Under this alternative, WS would not be involved in Wildlife Damage Management (WDM) to reduce predation to State and Federally listed species. No direct impacts would be experienced by any wildlife species or population as a result of WS operational control methods. Efforts by natural resource mangers and other entities to reduce or prevent predation could increase, which could result in impacts on nontarget species populations to an unknown degree. Impacts on nontarget species under this alternative could be the same, less than, or more than those of the proposed action depending on the level of effort expended by the natural resource managers.

**Alternative 2. Nonlethal Control before Lethal Control**

Under this alternative, WS would implement nonlethal control prior to the use of lethal methods. Impacts resulting from the implementation or recommendation of nonlethal control techniques and devices would be similar to Alternative 3; consequently, impacts associated with lethal control would be similar to Alternative 4. Overall, impacts of this alternative on nontarget species would be similar to Alternative 5.

**Alternative 3. Nonlethal Control Only**

Under this alternative, WS would only implement nonlethal control methods. Exclusion devices and live trap equipment used to capture feral cats and dogs would have minimal to no negative impacts on nontarget and T&E species. Nontarget species that are inadvertently captured in live traps (legholds, cage traps, and snares) would be released, if it is determined that it is safe to do so and if the animal is injury free. Nontarget risks are minimized by the selection of the appropriate trap size, pan tension, attractant (bait), and proper site selection. Frequent trap checks would further minimize risks to nontarget animals. To reduce the potential impacts to sea turtles, American crocodiles, shorebirds, and beach mice from WS activities, the placement and routine checking of trap and snare sets on the beach and/or primary and secondary dunes would be conducted during daylight hours, but before the temperature reached levels detrimental to the restrained animal. If nighttime operations are necessary, human presence would be kept to the minimum time necessary to conduct the operation. An exception to the time limitation would be to retrieve a captured animal. Risks associated with snares are greatest for animals that frequent the areas where snares are placed and travel along the paths of the target species. Nontarget risks could be further minimized by adjusting the size of the loop and the height of placement. Proper loop size and placement allows animals smaller than the target species to pass through or under the device unharmed. The use of break away locks and stops (device used to prevent a snare from choking an animal) will allow animals larger than the target species to break free of the device and nontarget animals to be released. Hazards to nontarget animals associated with the use of snares could range from minor injuries or potential death due to strangulation. Snare use by WS employees experienced in targeting and capturing specific animals will further minimize risks to nontarget animals. Observations during sea turtle nesting surveys indicate that humans speaking quietly in the vicinity do not disrupt turtle nesting behavior; however, movement does. Little information is available regarding impacts to colonial nesting birds and small mammals

5

from human presence on the dunes during nighttime hours. Human presence could disrupt or deter beach mice from leaving their burrows to forage. Continued human presence during nighttime hours could disrupt normal mouse behavior, cause undue stress, and lead to reduced overall health.

WS SOP's and mitigation measures, as described in 3.3, would be followed to help minimize potential impacts to nontarget and T&E species. The Florida WS program has captured a relatively low number of nontarget animals while conducting T&E species protection programs. Furthermore, no T&E species have been captured or injured by WS in Florida. See Section 2.2.3 for specific details.

**Alternative 4. Population Reduction (Lethal Control)**

Under this alternative, WS would implement lethal control methods without applying nonlethal methods. Lethal removal by shooting is nearly 100% selective for target species, thus no nontarget or T&E species are expected to be lethally removed as a result in WS utilizing selective shooting under this alternative. Ground shooting during nighttime hours could cause impacts to nesting or hatchling sea turtles or other T&E species from the use of lights to locate predators, or the presence of humans on the beach and/or primary or secondary dunes. Lights can inhibit female sea turtles from coming ashore to nest and can disorient turtle hatchlings as they emerge from the nests and crawl to the sea. Disorientation could prevent the hatchlings from reaching the sea, exposing them to dehydration and predation. Use of lights, during the night, outside of the nesting season would not cause problems for sea turtles or colonial nesting birds. Spotlights using red lens would lessen any potential impacts on T&E species during nesting season. Observations during sea turtle nesting surveys indicate that humans speaking quietly in the vicinity do not disrupt turtle nesting behavior; however, movement does. Little information is available regarding impacts to colonial nesting birds and small mammals from human presence on the dunes during nighttime hours. Human presence could disrupt or deter beach mice from leaving their burrows to forage. Continued human presence during nighttime hours could disrupt normal mouse behavior, cause undue stress, and lead to reduced overall health. Potential impacts associated with spotlights would be minimized by use of appropriate night vision equipment or red filtered spotlights. Human presence would be kept to the minimal time needed to accomplish the locating, shooting, and retrieval of predators. Impacts associated with firearm discharge and noise would be minimized through the use of air rifles and suppressed rifles, and the use of well trained personnel.

Nontarget animals that are inadvertently captured in live traps (legholds, cage traps, and snares) would be released if it is determined that it is safe to do so and if the animal is injury free. Nontarget risks are minimized by the selection of the appropriate trap size, use of pan tension devices, selection of the appropriate attractant (bait), and proper site selection. Frequent trap checks will further minimize risks to nontarget animals. To reduce the potential impacts to sea turtles, American crocodiles, shorebirds, and other protected species from WS activities, the placement and routine checking of trap and snare sets on the beach and/or primary and secondary dunes would be conducted during daylight hours, but before the temperature reaches levels detrimental to the trapped animal. If nighttime operations are necessary, human presence would be kept to the minimum time necessary to conduct the operation. An exception to the time limitation would be to retrieve a captured animal. Risks associated with snares are greatest for animals that frequent the areas where snares are placed and travel along the paths of the target species. Nontarget risks would be minimized by adjusting the size of the loop and the height of placement. Proper loop size and placement allows animals smaller than the target species to pass through or under the device unharmed. The use of break away locks and stops (device used to prevent a snare from choking an animal) will allow animals larger than the target species to break free of the device and nontarget animals to be released. Hazards to nontarget animals associated with the use of snares could range from minor injuries or potential death due to strangulation. Snare use by employees experienced in targeting and capturing specific animals will further minimize risks to nontarget animals.

WS SOP's and mitigation measures, as described in 3.3, would be followed to help minimize potential impacts to nontarget and T&E species. The Florida WS program has captured a relatively low number of nontarget animals while conducting T&E species protection programs. Furthermore, no T&E species have been captured or injured by WS in Florida. See Section 2.2.3 for specific details.

**Alternative 5. Integrated Wildlife Damage Management (Proposed Action)**

Under this alternative, WS would incorporate select components from Alternatives 2, 3, and 4 into its WDM program. Impacts resulting from the implementation or recommendation of nonlethal control techniques and devices would be similar to Alternative 3. The potential effects of lethal techniques would be similar to Alternative 4. Overall, impacts of control methods of this alternative on nontarget and T&E species would be similar to Alternative 2.

**4.1.4    Humaneness of Control Techniques**

**Alternative 1.  No Action**

Under this alternative, WS would not be involved in WDM to reduce predation to State and Federally listed species. No direct impacts would be experienced by any wildlife species or population as a result of WS operational control methods. Efforts by natural resource mangers and other entities to reduce or prevent predation could increase, potentially resulting in impacts on nontarget species populations to an unknown degree. Impacts on nontarget species under this alternative could be the same, less than, or more than those of the proposed action, depending on the level of effort expended by the natural resource managers.

**Alternative 2.  Nonlethal Control before Lethal Control**

Under this alternative, WS would be required to implement nonlethal methods prior to the implementation of lethal methods. Nonlethal methods could include live trapping and transporting feral/free-ranging cats and dogs to local animal shelters. Lethal methods, if implemented, would include shooting and live trapping followed by euthanasia. When performed by experienced professionals, shooting usually results in a quick death for the selected animal. WS personnel in Florida are experienced and professional in their use of control methods and implement these methods in the most humane manner possible. Mitigation measures and SOPs used to maximize humaneness were listed in Chapter 3.

7

Some segments of the public would view the shooting or killing an animal as inhumane. Persons or publics who view killing of any kind as inhumane would strongly oppose this alternative. Groups that are opposed to trapping and/or restraining of animals in traps and snares would considered this alternative inhumane. Overall, humanness of WDM under this alternative would be similar to Alternative 5.

**Alternative 3.  Nonlethal Control Only**

Under this alternative, WS would implement nonlethal control methods only. Nonlethal methods could include live trapping and transporting feral/free-ranging cats and dogs to local animal shelters.  WS personnel in Florida are experienced and professional in their use of control methods and use these methods in the most humane manner possible.  Mitigation measures and SOPs used to maximize humaneness were listed in Chapter 3.  Persons opposed to the live capturing and restraining of animals (i.e., traps and snares) would consider this alternative inhumane. Others that view lethal control of any kind as inhumane would most likely prefer this alternative to Alternatives 2, 4 and 5.

**Alternative 4.  Population Reduction (Lethal Control)**

Under this alternative, WS would implement lethal control methods without applying and considering nonlethal methods.  Lethal methods would generally be applied as a result of unsuccessful attempts by natural resource managers to alleviate predator damage through nonlethal methods.  Lethal methods would consist of selective shooting and live trapping followed by euthanasia. When performed by experienced professionals, shooting usually results in a quick death for the selected animal.  WS personnel in Florida are experienced and professional in their use of control methods and use these methods in the most humane manner possible.  Mitigation measures and SOPs used to maximize humaneness were listed in Chapter 3.

Some segments of the public would view the shooting or killing of an animal as inhumane. Persons or publics who view killing of any kind as inhumane would strongly oppose this alternative.  Groups that are opposed to trapping and/or restraining of animals in traps and snares would also considered this alternative inhumane.  Overall, humanness of WDM under this alternative would be similar to Alternatives 2 and 5.

**Alternative 5.  Integrated Wildlife Damage Management (Proposed Action)**

Under this alternative, WS would incorporate select components from Alternatives 2, 3, and 4 into its WDM program.  Humaneness would be of the same level as that in Alternatives 2 and 4.

**4.1.5    Effects of Control Methods on Human Health and Safety**

**Alternative 1.  No Action**

Under this alternative, WS would not be involved in Wildlife Damage Management (WDM) to reduce predation to State and Federally listed species. Therefore, WS damage control activities and methods would have no direct impact on human health and safety.

Risks to human safety from WS's use of firearms and trapping devices would be alleviated because no such use would occur. However, increased use of firearms and traps by less experienced and trained private individuals would probably occur. WS would not provide assistance to private individuals in the safe and proper use of WDM control devices. Risks to human safety could increase under this alternative, but probably not significantly.

**Alternative 2. Nonlethal Control before Lethal Control**

Under this alternative, WS would be required to implement nonlethal methods prior to the implementation of lethal methods. WDM methods that might raise safety concerns include shooting with firearms and the use of traps and snares. Firearms are only used by WS personnel who are experienced in the safe handling and operation of such devices. WS personnel receive firearms safety training on an annual basis to keep them aware of safety concerns. The Florida WS Program has not had any accidents involving the use of firearms or traps and snares in which a member of the public was harmed. Mitigation measures and SOPs used to maximize safe use of control methods were listed in Chapter 3. A formal risk assessment of WS's operational management methods found that risks to human safety were low (USDA 1994, Appendix P). Therefore, no significant impacts on human safety from WS's use of these methods is expected.

**Alternative 3. Nonlethal Control Only**

Under this alternative, WS would implement and recommend nonlethal control methods only. WDM methods that might raise safety concerns include the use of traps and snares for the live capture and transport of feral/free-ranging cats and dogs to local animal shelters. WS personnel receive safety training on an annual basis to keep them aware of safety concerns. The Florida WS Program has not had any accidents involving the use of traps and snares in which a member of the public was harmed. Mitigation measures and SOPs used to maximize safe use of control methods were listed in Chapter 3. A formal risk assessment of WS's operational management methods found that risks to human safety were low (USDA 1994, Appendix P). Therefore, no significant impacts on human safety from WS's use of these methods is expected.

**Alternative 4. Population Reduction (Lethal Control)**

Under this alternative, WS would implement lethal control methods without applying or considering any nonlethal methods. Lethal methods would generally be applied as a result of unsuccessful attempts by natural resource managers to alleviate predator damage through nonlethal methods. WDM methods that might raise safety concerns include shooting with firearms and the use of traps and snares. Firearms are only used by WS personnel who are experienced in the safe handling and operation of such devices. WS personnel receive firearms safety training on an annual basis to keep them aware of safety concerns. The Florida WS Program has not had any accidents involving the use of firearms or traps and snares in which a member of the public was harmed. Mitigation measures and SOPs used to maximize safe use of control methods were listed in Chapter 3. A formal risk assessment of WS's operational management methods found that risks to human safety were low (USDA 1994, Appendix P). Therefore, no significant impacts on human safety from WS's use of these methods is expected.

**Alternative 5. Integrated Wildlife Damage Management (Proposed Action)**

Under this alternative, WS would incorporate select components from Alternatives 2, 3, and 4 into its WDM program. Potential impacts associated with this alternative would be similar to those in Alternatives 2 and 4.

**4.1.6    Effects on the Aesthetic Values of Targeted Species and Protected T&E Species**

**Alternative 1. No Action**

Under this alternative, WS would not conduct any lethal or nonlethal Wildlife Damage Management (WDM) activities towards the protection of the said species and groups. Some people and/or groups who oppose any wildlife damage control by government agencies or other groups and individuals would support this alternative. People or groups who have affectionate bonds with individual animals or animals in general, would not be affected by WS activities as stated in this alternative. Conversely, large segments of the public who value T&E species would be impacted negatively because of the continued high level of predation on these listed species and their continued reduction and potential extinction. However, it is likely that other natural resource managing agencies would conduct similar WDM on properties with this concern, resulting in impacts similar to those addressed in the WS Proposed Action.

**Alternative 2. Nonlethal Control before Lethal Control**

Under this alternative, WS would conduct nonlethal control methods prior to carrying out lethal control. It is important to note, that prior to WS involvement, most agencies and citizen groups involved in the management of T&E species have exhausted the use of nonlethal control methods. Some people have expressed opposition to the killing of any animals during WDM activities. Under this alternative some lethal control of predators could occur and these persons would continue to be opposed. However, many persons who voice opposition have no direct connection or opportunity to view or enjoy the particular animals that would be killed by WS's lethal control activities. Lethal control actions would generally be restricted to local sites and to small, insubstantial percentages of the overall population. Therefore, the species subjected to limited lethal control actions would remain common and abundant; therefore, these animals (as a species) would still be available for viewing by persons with that interest. Some segments of the public are concerned about the welfare and potential impacts to feral/free-ranging cats and dogs. These publics would likely favor this alternative and Alternatives 3, 4, and 5, since these animals would be taken to local animal shelters for further assistance in their well being.

The requirement for WS to implement nonlethal methods before lethal control would prolong predation impacts and would be detrimental to T&E species. Publics concerned with T&E protection would be negativity impacted because of the continued level of predation sustained by these species. Overall, impacts of this alternative on target species would be similar to Alternatives 4 and 5; conversely, the negative impacts to protected T&E species would be greater than Alternatives 4 and 5 and similar to Alternatives 1 and 3.

**Alternative 3. Nonlethal Control Only**

Under this alternative, WS would implement and recommend nonlethal control methods only. No impacts to predator species would be expected as the direct result of WS operations, except that feral cats and dogs would be captured and transported to local animal shelters. Persons whom are concerned with the welfare and potential impacts to feral/free-ranging cats and dogs would likely favor this alternative and Alternatives 2, 4, and 5, since these feral animals would be taken to animal shelters for further assistance in their well being.

The requirement for WS to implement nonlethal methods would prolong predation impacts and would be detrimental to T&E species. Publics concerned with T&E protection would be negativity impacted because of the continued level of predation sustained by these species. Overall, impacts of this alternative on target species would be slightly greater than Alternative 1 and less than Alternatives 2, 4 and 5. Negative impacts to protected T&E species would be greater than Alternatives 4 and 5 and similar to Alternatives 1 and 2.

**Alternative 4. Population Reduction (Lethal Control)**

Under this alternative, WS would implement and recommend lethal control methods without applying or considering nonlethal methods. Some people have expressed opposition to the killing of any animals during WDM activities. Under this alternative some lethal control of predators could occur and these persons would continue to be opposed. However, many persons who voice opposition have no direct connection or opportunity to view or enjoy the particular animals that would be killed by WS's lethal control activities. Lethal control actions would generally be restricted to local sites and to small, insubstantial percentages of the overall population. Therefore, the species subjected to limited lethal control actions would remain common and abundant; therefore, these animals (as a species) would still be available for viewing by persons with that interest. Some segments of the public are concerned about the welfare and potential impacts to feral/free-ranging cats and dogs. These publics would likely favor this alternative and Alternatives 3, 4, and 5, since these animals would be taken to local animal shelters for further assistance in their well being.

Publics concerned with T&E protection would likely favor this alternative because predation rates to T&E species would be reduced under this alternative; therefore, increasing the likelihood of the continued survival of the T&E species proposed for protection from predation. Overall, impacts of this alternative on target species would be similar to Alternatives 2 and 5. Negative impacts to the protected T&E species would be less than Alternatives 1, 2 and 3 and similar to Alternative 5.

**Alternative 5. Integrated Wildlife Damage Management (Proposed Action)**

Under this alternative, WS would incorporate select components from Alternatives 2, 3, and 4 into its WDM program. Potential impacts associated with this alternative would be similar to those in Alternative 4.

## 4.2  CUMULATIVE IMPACTS

No significant or cumulative adverse environmental consequences resulting from the proposed action are anticipated (Table 4-4). Control activities will not negatively impact other protected flora or fauna. Beneficial impacts are expected to be increased nesting success of the loggerhead, green, hawksbill, Kemp's ridley, leatherback sea turtles, and American crocodile; and reduced predation threats to the Perdido Key beach mouse, Chocatawhatchee beach mouse, Key Largo cotton mouse, Anastasia Island beach mouse, St. Andrews beach mouse, Santa Rosa Island beach mouse, Lower Keys marsh rabbit, silver rice rat, Key Largo woodrat, and Key Largo cotton mouse; and increased habitat quality and nesting success for snowy plover, piping plover, American oystercatcher, black skimmer, roseate tern, and least tern.

Federal and State wildlife agencies were contacted concerning the Proposed Action and reviewed this document concerning any potentially negative impacts to the environment.

This approach has previously been used effectively by WS to reduce predation losses involving > 30 threatened or endangered species projects in California, Alaska, Nebraska, and Hawaii, during fiscal years 1995-2000. WS would conduct management activities as needed, to remove predating/damage causing species. Natural resource managers and their personnel would continue using exclusion devices.

To assure that visitors will not be in the areas of predator control work during nighttime hours, additional precautions may be taken besides the precautions discussed in Alternative 3. Signs would be placed along the beach and/or on trails where work is being conducted, instructing visitors to stay out of the area. If visitors are seen in the work area, they will be asked to leave and remain out of the work area.

Removal of predators from concerned areas will resolve the immediate problem; however, over time, other predators will move in from surrounding areas and replace the ones taken. These immigrants may not be trained to exploit sea turtle nests, but since it is a learned behavior, they will likely become nest predators. Also, coyotes, foxes, and raccoons are natural predators of rodents and birds, and any of these predators within the concerned areas would be potential threats to T&E species. Because of these factors, any work plan for a predator damage management project will have to include long-term plans, using the integrated wildlife damage management approach outlined in this EA. All populations of the listed species addressed in this EA are entirely dependent on very limited and dwindling coastal habitats for their survival, and face the possibility of extinction. Consequently, it is essential that immediate actions be taken to reduce the likelihood of extinction.

No threatened or endangered species or critical habitat would be adversely impacted by the proposed action. Therefore, WS with concurrence from the USFWS, has determined that the proposed action would not likely adversely affect any species protected under the U.S. Endangered Species Act.

Table 4-4. Summary of the potential effects of the Alternatives as it pertains to the identified Issues. Potential effects include both positive and negative, when applicable.

| ISSUES | ALTERNATIVE 1 NO ACTION | ALTERNATIVE 2 NONLETHAL CONTROL BEFORE LETHAL CONTROL | ALTERNATIVE 3 NONLETHAL CONTROL ONLY | ALTERNATIVE 4 LETHAL CONTROL ONLY | ALTERNATIVE 5 INTEGRATED WILDLIFE DAMAGE MANAGEMENT (PROPOSED ACTION) |
|---|---|---|---|---|---|
| EFFECTS OF PREDATION ON RESOURCES PROTECTED, INCLUDING NATIVE WILDLIFE AND PLANTS | Moderate to High Impact on all T&E species proposed for protection. Feral hog and predator damage would continue at the current rate. | Moderate to High Impact on all T&E species proposed for protection. Predation would gradually be alleviated as the use of lethal methods became available; High initial Impact to T&E species because of continued predation rates. | Moderate to High Impact on all T&E species proposed for protection. This alternative would maintain or increase the current impact of predation on many of the T&E species listed in this EA. | Low Impact to all T&E species proposed for protection. Low to No Negative Impact to other native plant and animal species. This alternative would alleviate predation for most T&E species under protection. | Low Impact to all T&E species proposed for protection. Low to No Negative Impact to other native plant and animal species. This alternative would alleviate predation for most T&E species under protection. |
| EFFECTS ON TARGET SPECIES POPULATIONS | No Impact would occur from WS wildlife damage management. | Low Impact to target species; impact would be localized in nature. | No Impact would occur from WS wildlife damage management. | Low Impact - but greater than Alternatives 1 & 3; target species impact would be localized in nature. | Low Impact - but greater than Alternatives 1 & 3. All impacts to target species would be localized in nature. |
| EFFECTS OF CONTROL METHODS ON NONTARGET SPECIES POPULATIONS, INCLUDING T&E SPECIES | No direct Impact would be observed with any nontarget or T&E species as a result of WS operations. | Low Impact, but greater than Alternative 1 & 3; Indirect Impact would be Higher for T&E species because of continued predation. After the implementation of lethal control methods, High to Moderate positive Impact would be seen for T&E species, due to alleviation of predation. All traps proposed for use are live capture devices; any nontarget animal captured will be released, whenever possible. | No direct Impact would be observed with any nontarget or T&E species as a result of implementation of nonlethal methods by WS; High to Moderate Indirect Impact to T&E species, as a result of continued high predation rates. | Low Impact, but greater than Alternative 1 & 3; High to Moderate - positive Impact to T&E species, due to alleviation of predation. All traps proposed for use are live capture devices; any nontarget animal captured will be released, whenever possible. | Low Impact, but greater than Alternative 1 & 3; High to Moderate - positive Impact to T&E species, due to alleviation of predation. All traps proposed for use are live capture devices; any nontarget animal captured will be released, whenever possible. |

Table 4-4. Continued.

| ISSUES | ALTERNATIVE 1. NO ACTION | ALTERNATIVE 2. NONLETHAL CONTROL BEFORE LETHAL CONTROL | ALTERNATIVE 3. NONLETHAL CONTROL ONLY | ALTERNATIVE 4. LETHAL CONTROL ONLY | ALTERNATIVE 5. INTEGRATED WILDLIFE DAMAGE MANAGEMENT (PROPOSED ACTION) |
|---|---|---|---|---|---|
| HUMANENESS OF CONTROL TECHNIQUES | No Impact - Under this Action none of the proposed species to be controlled would be managed by WS. | Low to Moderate Impact - greater than Alternatives 1 & 3, but as humane as possible with the available resources and technologies. | Low Impact - less than Alternatives 2, 4, & 5. Under this action none of the proposed species to be controlled would be managed by WS. | Low to Moderate Impact - greater than Alternatives 1 & 3, but as humane as possible with the available resources and technologies. | Low to Moderate Impact - greater than Alternatives 1 & 3, but as humane as possible with the available resources and technologies. |
| EFFECTS OF CONTROL METHODS ON HUMAN HEALTH AND SAFETY | No Impact - no potential Human Health and Safety issues would be created by the WS operational program. | Low Risk or Impact - but greater than Alternatives 1 & 3. | Low Risk or Impact - no potential Human Health and Safety issues would be created by the WS operational program. | Low Risk or Impact - as the result of WS operations, but greater than Alternatives 1 & 3. | Low Risk or Impact - as the result of WS operations, but greater than Alternatives 1 & 3. |
| EFFECTS ON THE AESTHETICS VALUES OF TARGETED SPECIES AND PROTECTED T&E SPECIES | No Impact - for the species proposed for control; High Impact for T&E species. The aesthetics of the T&E species proposed for protection would be greatly affected by this Action. | Low Impact - predators will not be eliminated from a system and most of the predators proposed for control are nocturnal, thus seldom observed by most people. Potential, High initial Impact for T&E species until the implementation of lethal control techniques. | No Impact - for target species; High to Moderate Impact - in respect to the aesthetics of the T&E species proposed for protection, as a result of continued predation. | Low Impact - predators will not be eliminated from a system and most of the predators proposed for control are nocturnal, thus seldom observed by most people. High positive Impact for T&E species proposed for protection, due to alleviation of predation. | Low Impact - predators will not be eliminated from a system and most of the predators proposed for control are nocturnal, thus seldom observed by most people. High positive Impact for T&E species proposed for protection, due to alleviation of predation. |

## CHAPTER 5: LIST OF PREPARERS AND PERSONS CONSULTED

### PREPARERS

Lawrence J. Brashears, Jr.    USDA, APHIS, WS - District Supervisor
Bernice U. Constantin    USDA, APHIS, WS - State Director
David S. Reinhold    USDA, APHIS, WS - Environmental Management Coordinator

### CONSULTATIONS

Carmen Simonton    USFWS, Atlanta, GA
Brian Millsap    FFWCC, Tallahassee, FL
James E. Moyers    St. Joe Timberlands Company, Port St. Joe, FL
Sandra MacPherson    USFWS, Jacksonville, FL
Lorna Patrick    USFWS, Panama City, FL
John Bente    FDEP, Panama City, FL
Guy Connolly    USDA, APHIS, DWRC, Lakewood, CO
Joe Mitchell    FDEP, Saint Joseph Peninsula State Park, FL
Jeff Gore    FFWCC, Panama City, FL
Allen Foley    FFWCC/FMRI, St. Petersburg, FL
Kerri Powell    FFWCC/FMRI, St. Petersburg, FL
Carl Petrick    DOD, Eglin AFB, Ft. Walton, FL
Dennis Teague    DOD, Eglin AFB, Ft. Walton, FL
Mark Nicholas    NPS, Gulf Islands National Seashore, Pensacola, FL
Richard Crossett    FFWCC, Quincy, FL
Terry J. Doyle    USFWS, Ten Thousand Islands NWR, Naples, FL
Ben Nottingham    USFWS, Florida Panther NWR, Naples, FL
Ryan M. Noel    USFWS, Hobe Sound NWR, Hobe Sound, FL
Mark W. Nelson    FDEP, Jonathan Dickinson State Park, Hobe Sound, FL
John R. Griner    FDEP, St. Lucie Inlet/Seabranch State Preserves, Hobe Sound, FL
Skip Snow    NPS, Everglades National Park, Homestead, FL
Glen Dodson    NPS, Everglades National Park, Everglades City, FL
Kiefer Gier    NPS, Everglades National Park, Everglades City, FL

## APPENDIX A. BIBLIOGRAPHY AND LITERATURE CITED

Bailey, J.A. 1984. Principles of wildlife management. John Wiley and Sons, New York, New York. 373pp.

Balser, D. S., D. H. Dill, and H. K. Nelson. 1968. Effect of predator reduction on waterfowl nesting success. J. Wildl. Manage. 32:669-682.

Bekoff, M. 1977. Canis latrans. The American Society of Mammalogist. Mammalian Species No. 79:1-9.

Berryman, J.H. 1991. Animal damage management: responsibilities of various agencies and the need for coordination and support. Proc. East Wild. Damage Control Conf. 5:12-14.

Cunningham, V. D. and R. D. Dunford. 1970. Recent coyote record from Florida. Quart. Jour. Florida Acad. Sci. 33:279-280.

Davis, G. E. and M. C. Whiting. 1977. Loggerhead sea turtle nesting in Everglades National Park, Florida, USA. Herpetologica 33:18-28.

Day, D. 1981. The Doomsday Book of Animals. London Editions Limited, 70 Old Compton Street, London. 288 pp.

DeBenedetti, S. H. 1986. Management of feral pigs at Pinnacles National Monument: Why and How. In Proceedings of the conference on the conservation and management of rare and endangered plants. California Native Plant Society. Sacramento, CA.

Florida Fish and Wildlife Conservation Commission. 1999. Ground colonial shorebird colonies occupied in 1999 and associated predators. (unpubl. rep.)

Frost, C. C. 1993. Four centuries of changing landscape patterns in the longleaf pine ecosystem. Pages 17-37 in S. M. Hermann, ed. The longleaf pine ecosystem: Ecology, restoration, and management. Proceedings 18th Tall Timbers Fire Ecology Conference: Tallahassee, FL.

Garmestani, A. S. 1997. Sea Turtle Nesting in the Ten Thousand Islands of Florida, Chapter 4. M.S. Thesis. University of Florida, Gainesville, Florida.

Gore, J. A. and M. J. Kinnison. 1991. Hatching success in roof and ground colonies of least terns. Condor 93:759-762.

Greenwood, R. J. 1986. Influence of striped skunk removal on upland duck nest success in North Dakota. Wildl. Soc. Bull. 14:6-11.

Grover, P. B. 1979. Habitat requirements of Charadriiform birds nesting on salt flats at Salt Plains National Wildlife Refuge. M.S. Thesis, Oklahoma State Univ., Stillwater, OK.

_____ and F. L. Knopf. 1982. Habitat requirements and breeding success of Charadriiform birds nesting at Salt Plains National Wildlife Refuge, Oklahoma. J. Field Ornithol. 53:139-148.

1

Gruver, K. S., R. L. Philips, and E. S. Williams. 1996. Leg injuries to coyotes captured in standard and modified Soft Catch traps. Proc. Vertebr. Pest. Conf. 17.

Jurek, R. M. 1994. A bibliography of feral, stray, and free-ranging domestic cats in relation to wildlife conservation. Calif. Dep. of Fish and Game, Nongame Bird and Mammal Program Rep. 94-5. 24 pp.

Harrison, G. H. 1992. Is there a killer in your house? National Wildlife, October/November Issue: 10-15.

Holler, N. R. 1992. Choctawhatchee beach mouse. Pages 76-86 in S. R. Humphrey, ed., Rare and endangered biota of Florida, Vol. 1. Mammals. University Press of Florida. Tallahassee, FL. 392 pp.

Holliman, D. C. 1983. Status and habitat of Alabama gulf beach mice, *Peromyscus polionotus amobates* and *P. p. trissyllepsis*. Northeast Gulf Sci. 6: 121-129.

Hubbs, E. L. 1951. Food habits of feral house cats in the Sacramento Valley. Calif. Fish and Game 37: 177-189.

Johnson, A. S. 1970. Biology of the raccoon in Alabama. Auburn Univ. Exper. Stn. Bull. 402. 148 pp.

Johnson, E. F. and E. L. Rauber. 1970. Control of raccoons with rodenticides. Pro. Annu. SE Conf. Fish and Wildl. Agencies 24:277-281.

Kirsch, E. M. 1996. Habitat selection and productivity of least terns on the lower Platte River, Nebraska. Wildl. Monogr. 132:1-48.

Leopold, A.S. 1933. Game management. Charles Scribner & Sons, New York, NY. 481 pp.

Lipscomb, D. J. 1989. Impacts of feral hogs on longleaf pine regeneration. Southern J. of Applied Forestry 13(4):177-181.

MacIvor, L. H., S. M. Melvin, and C. R. Griffin. 1990. Effects of research activity on piping plover nest predation. J. Wildl. Manage. 54:443-447.

Massey, B. W. 1971. A breeding study of the California least tern, 1971. Admin. Rep. 71-9, Wildl. Manage. Branch, Calif. Dept. Fish and Game, Helen, MT.

_____ and J. L. Atwood. 1979. Application of ecological information to habitat management for the California least tern. Prog. Rep. 1, U. S. Fish and Wildlife Serv., Laguna Niguel, CA.

Means, D. B. 1999. *Desmognathus* auriculatus. Pages 10-11 in Michael Lanoo, ed. Status and Conservation of U.S. Amphibians. Declining Amphibians Task Force Publ. No. 1.

Minsky, D. 1980. Preventing fox predation at a least tern colony with an electric fence. J. Field Ornithol. 51:180-181.

Moyers, J. E. 1996. Food habits of gulf coast subspecies of beach mice (*Peromyscus polionotus spp.*). M.S. Thesis. Auburn University, Auburn, Alabama. 83 pp.

National Academy of Sciences. 1990. Biotic sources of mortality. Pages 62-64 in Decline of the sea turtle: causes and prevention - Committee on Sea Turtle Conservation, Board on Environmental Studies and Toxicology, Board on Biology, Commission on Life Sciences, National Research Council. National Academy Press. Washington, D.C.

Paradiso, J. L. 1968. Canids recently collected in East Texas, with comments on the taxonomy of the red wolf. Amer. Midland Nat. 80:529-534.

Parker, G.R. 1995. Eastern coyotes. Nimbus Publishing Limited, Halifax, N.S. 254pp.

Phillips, R. L., K. S. Gruver, and E. S. Williams. 1996. Leg injuries to coyotes captured in three types of foothold traps. Wildl. Soc. Bull. 24(2): 260-263.

Printiss, D. and D. Hipes. 1999. Rare amphibian and reptile survey of Eglin Air Force Base, Florida. Florida Natural Areas Inventory Final Report. Tallahassee, FL.

Raymond, P. W. 1984. Sea Turtle hatchling disorientation and beach front lighting. A review of the problem and potential solutions. The Center for Environmental Education. Sea Turtle Rescue Fund. Washington, D. C. 72 pp.

Seabrook W. 1989. Feral cats (Felis catus) as predators of hatchling green turtles (Chelonia mydas). J. Zool., Lond. 219: 83-88.

Singer, F. J., W. T. Swank, and E. E. C. Clebsch. 1982. Some ecosystem responses to European wild boar rooting in a deciduous forest. Research/Resources Management Report No. 54. USDI, National Park Serv.: Atlanta, GA.

Southwest Florida Water Management District. 1996. Feral hog hunt is a squealin' success. Water Management Monthly 6(1):2.

Speake, D. W. 1980. Predation on wild turkeys in Alabama. Proc. Fourth Natl. Wild Turkey Symp. 4:86-101.

_____, R. Metzler, and J. McGlincy. 1985. Mortality of wild turkey poults in Northern Alabama. J. Wildl. Manage. 49:472-474.

_____, L. H. Barwick, H. O. Hillestad, and W. Stickney. 1969. Some characteristics of an expanding turkey population. Proc. Annu. Conf. SE Assoc. Fish and Wildl. Agencies 23:46-58.

Stallcup, R. 1992. Cats: a heavy toll? Bird Watcher's Digest. March/April Issue: 95.

Slate, D. A., R. Owens, G. Connolly, and G. Simmons. 1992. Decision making for wildlife damage management. Trans. North Am.. Wildl. Nat. Resour. Conf. 57:51-62.

Stancyk, S. E. 1979. Non-human predators of sea turtles and their control. Pages 139-152 in K. A. Bjornal, ed., Biology and conservation of sea turtles. Proceedings of the World Conference on Sea Turtle Conservation. Washington, D. C.

Thompson, R. L. 1977. Feral hogs on National Wildlife Refuges. Pages 11-16 in G. W. Wood, ed., Research and management of wild hog populations: Proceedings of a Symposium. Georgetown, S. C. 113 pp.

Till, J. A., and F. F. Knowlton. 1983. Efficacy of denning in alleviating coyote depredations on domestic sheep. J. Wildl. Manage. 47:1018-1025.

Trautman, C. G., L. F. Fredrickson, and A. V. Carter. 1974. Relationship of red foxes and other predators to populations of ring-necked pheasants and other prey, South Dakota. Trans. North Am. Wildl. Nat. Resour. Conf. 39:241-252.

USDA. 1989. Strategic Plan. U.S. Dept. Agric., Anim. Plant Health Inspection Serv., Anim. Damage Control, Operational Support Staff, Riverdale, MD.

_____. 1991. Wild pigs: Hidden danger for farmers and hunters. Agri. Infor. Bull. No. 620. 6 pp.

_____. 1994. Animal Damage Control program final environmental impact statement. Vol. 1-3. Animal Damage Control. Hyattsville, MD.

U. S. Fish and Wildlife Service. 1985. Endangered and threatened wildlife and plants; determination of endangered status and critical habitat for three beach mice; final rule. Federal Register 50(109): 23872-23889.

_____. 1993. Recovery plan for the Anastasia Island and Southeastern Beach Mouse. Atlanta, GA. 3 pp.

_____. 1999. South Florida multi-species recovery plan. Atlanta, Georgia. 2172 pp.

Van't Woudt, B. D. 1990. Roaming, stray, and feral domestic cats and dogs as wildlife problems. Proc. 14th Vert. Pest Conf.: 291-295.

Wildlife Services. 1999. Environmental Assessment for Predator damage management in Nebraska for the protection of livestock, wildlife, property and public health and safety. USDA, APHIS, WS. Nebraska..

Wildlife Society, The. 1990. Conservation policies of The Wildlife Society. The Wildlife Society. Washington, D.C. 20 pp.

Witherington, B. E. 1991. Influences of artificial lighting on the seaward orientation of hatchling loggerhead turtles. *(Caretta caretta)*. Biological Conservation 55:139-149.

_____. 1995. Hatchling orientation. A Summary. Pages 577-578 in K. A. Bjorndal, ed., Biology and Conservation of sea turtles. Revised Edition. Smithsonian Institution Press. Washington D.C.

_____ and R. E. Martin. 1996. Understanding, assessing, and resolving light-pollution problems on sea turtle nesting beaches. FMRI Tech. Rep. TR-2. Florida Marine Research Institute, St. Petersburg, Florida. 73p.

Yamalis, H. W. and T. J. Doyle. 1999. Sea Turtle Nesting Activity in Florida's Ten Thousand Islands from 1991 to 1999. USFWS, Ten Thousand Islands National Wildlife Refuge, Naples, Florida. (unpubl. rep.).

# APPENDIX B. GLOSSARY

**Abundance** - The number of individuals of a species in a given unit of area.

**Animal Behavior Modification** - The use of scare tactics/devises (i.e., electronic distress sounds, propane exploders, pyrotechnics, lights, scarecrows, etc.) to deter or repel animals that cause damage to resources or property or threaten human health and safety.

**Animal Rights** - A philosophical and political position that animals have inherent rights comparable to those of humans.

**Animal Welfare** - Concern for the well-being of individual animals, unrelated to the perceived rights of the animal or the ecological dynamics of the species.

**Canid** - A coyote, dog, fox, wolf or other member of the dog (Canidae) family.

**Carnivore** - A species that primarily eats meat (member of the Order Carnivora).

**Confirmed Losses** - Wildlife-caused losses or damages verified by USDA-WS. These figures usually represent a fraction of the total losses.

**Corrective Damage Management** - Management actions applied when damage is occurring or after it has occurred.

**Denning/Den Hunting** - The process of locating predator (primarily coyote) burrows and destroying the pups. The adult predator may also be killed.

**Depredating Species** - An animal species causing damage to, or loss of crops, livestock, other agricultural or natural resources, property, or wildlife.

**Depredation** - The act of killing, damaging, or consuming animals, crops, other agricultural or natural resources, property, or wildlife.

**Direct Control** - Administration or supervision of wildlife damage management by WS, often involving direct intervention to capture depredating animals.

**Endangered Species** - Federal designation for any species or population that is in danger of extinction throughout all or a significant portion of its range.

**Environment** - The conditions, influences, or forces that affect or modify an organism or and ecological community and ultimately determine its form and survival.

**Environmental Assessment** - An analysis of the impacts of a planned action to the human environment to determine the significance of that action and whether an EIS is needed.

**Environmental Impact Statement** - A document prepared by a federal agency to analyze the anticipated environmental effects of a planned action or development, compiled with formal examination of options and risks.

**Eradication** - Elimination of a specific wildlife species, generally considered pests, from designated areas.

**Exotic (Nonnative) Species** - Any plant or animal that is not native to an area; species transplanted by humans that are native to other areas of a county, state, or other parts of a country or species introduced from other countries.

**Feral (Nonnative) Wildlife Species** - Generally, any animal commonly domesticated by humans that is no longer dependent on humans to survive and living in the wild (i.e., escaped livestock, poultry, fowl, dogs, cats, etc.).

1

**Habitat** - An environment that provides the requirements (i.e., food, water, shelter, and space) essential for the development and sustained existence of a species.

**Habitat Modification/Management** - Protection, destruction, or modification of a habitat to maintain, increase, or decrease its ability to produce, support, or attract designated wildlife species

**Harvest or Kill Data** - An estimate of the number of animals removed from a population by humans.

**Humaneness** - The perception of compassion, sympathy, or consideration for animals from the viewpoint of humans.

**Integrated Pest Management** - The procedure of integrating, applying, and assessing practical pest management methods while minimizing potential harmful effects to humans, nontarget species, and the environment. Often several different techniques are incorporated into a management program (i.e., cultural, exclusion, lethal and nonlethal methods, etc.).

**Integrated Wildlife Damage Management** - See *Integrated Pest Management*. The IPM approach applied to the objective of managing wildlife damage rather than pest animal populations. Often several different techniques are incorporated into a management program (i.e., cultural, exclusion, lethal and nonlethal methods, etc.).

**Lethal Management Methods/Techniques** - Wildlife damage management methods that result in the death of targeted animals (e.g., ground calling and shooting, trapping, denning, etc.).

**Local Population** - The population within an immediate specified geographical area.

**Long-term** - An action, trend, or impact that affects the potential of an event over an extended period of time.

**Magnitude** - Criteria used in this EA to evaluate the significance of impacts on species abundance. Magnitude refers to the number of animals removed in relation to their abundance.

**Nonlethal Control Methods/Techniques** - Wildlife damage management methods or techniques that do not result in the death of targeted animals ( e.g., live traps, repellents, pyrotechnics, fences, etc.).

**Nontarget Species/Animals** - An animal species or local population that is inadvertently captured, killed, or injured during wildlife damage management and is not the targeted species/animal.

**Offending Animal/Species** - The individual animal(s) within a specified area causing damage to property, public health and safety, wildlife, natural resources, or to agricultural resources.

**Omnivore/Omnivorous** - An animal that eats both plant and animal matter; a generalist, opportunistic feeder that eats whatever is available.

**Pesticide** - A toxic chemical substance used to control pest animals.

**Population** - A group of organisms of the same species that occupies a particular area.

**Predator** - An animal that kills and consumes another animal.

**Preventive Damage Management** - Management applied before damage begins.

**Prey** - An animal that is killed and consumer by a predator.

**Pyrotechnics** - Specialize fireworks used to frighten wildlife.

**Repellent** - A substance with taste, odor, or tactile properties that discourages specific animals or species from using a food or place.

**Requester** - Individual(s) or agency(ies) that request wildlife damage management assistance from WS.

**Selectivity** - Damage management methods that affect the specific animals or species responsible for causing damage without adversely affecting other species.

**Short-term** - An action, trend, or impact that does not have long lasting affects to the reproductive or survival capabilities of a species.

**Significant Impact** - An impact that will cause important positive or negative consequences to man and his environment.

**Take** - The capture or killing of an animal.

**Target Species/Animal/Population** - An animal, species, or population at which wildlife damage management is directed.

**Technical Assistance** - Advise, recommendations, information, demonstrations, and materials provided to others for managing wildlife damage problems.

**Threatened Species** - Federal designation for a species or population that is likely to become an endangered species within the foreseeable future throughout all or a significant portion of its range.

**Toxicant** - A poison or poisonous substance.

**Unconfirmed Losses** - Losses or damage reported by resource owners or managers, but not verified by WS.

**Wildlife** - Any wild mammal, bird, reptile, or amphibian.

**Wildlife Damage Management** - Actions directed toward resolving livestock or wildlife predation, protecting property, or safegaurding public health and safety in a coordinated, managed program.

**Work Plan** - A management plan developed jointly by WS and other federal, state, individuals, or other private entities specifying when, where, how, and under what constraints wildlife damage management will be conducted. Work plans generally include a map showing areas designated for planned control, restricted control, no control, and special protection.

# DECISION
## AND
## FINDING OF NO SIGNIFICANT IMPACT

**Management of Predation Losses to State and Federally
Endangered, Threatened, and Species of Special Concern; and
Feral Hog Management
to Protect Other State and Federally Endangered, Threatened,
Species of Special Concern, and
Candidate Species of Fauna and Flora
in the State of Florida**

The U.S. Department of Agriculture (USDA), Animal and Plant Health Inspection Service (APHIS), Wildlife Services (WS) program responds to requests for assistance from individuals, organizations and agencies experiencing damage caused by wildlife in Florida. WS has prepared an environmental assessment (EA) that analyzes alternatives for managing predation losses to state and federally endangered, threatened, species of special concern, and candidate species of plants and animals in the state of Florida. APHIS procedures for implementing the National Environmental Policy Act (NEPA) allows for the categorical exclusion of individual wildlife damage management actions (7 CFR 372.5(c), 60 Fed. Reg. 6000-6003, 1995). However, to properly address WS involvement in this action statewide, an EA was prepared to facilitate planning, interagency coordination, and the streamlining of program management, and to clearly communicate with the public the analysis of cumulative impacts. The pre-decisional EA released by WS in August 2001, documented the need for assisting natural resource managers in reducing predation losses to state and federally listed species in Florida and assessed potential impacts of various alternatives for responding to predation issues involving listed species. Comments from the public involvement process were reviewed for substantial issues and alternatives which were considered in developing this decision. The EA is tiered to the programmatic Environmental Impact Statement (EIS) for the Wildlife Services Program[1] (USDA 1997).

WS's proposed action was to implement an integrated wildlife damage management program that would include education and non-lethal and lethal methods to reduce predation losses to listed species throughout the State of Florida and to incorporate WS's current technical assistance approach to managing listed species and predator conflicts. Direct control assistance will only take place after a request for services has been received and where permission has been granted by private landowner or government manager. Based on the analysis in the EA, I have determined that there will not be a significant impact, individually or cumulatively, on the quality of the human environment from implementing the proposed action, and that the action does not constitute a major federal action significantly affecting the quality of the human environment.

---

[1] USDA (U.S. Department of Agriculture), Animal and Plant Health Inspection Service (APHIS), Animal Damage Control (ADC). 1997 (revised). Animal Damage Control Program, Final Environmental Impact Statement. Anim. Plant Health Inspection Serv., Anim. Damage Control. Hyattsville, MD. Volume 1, 2 & 3.

## Public Involvement

The pre-decisional EA was prepared and released to the public for a 30-day comment period by a legal notice in the Tampa Tribune, Tallahassee Democrat, Miami Herald, and The Florida Times Union (Jacksonville) on August 26, 2001. The pre-decisional EA was also mailed directly to agencies, organizations, and individuals with probable interest in the proposed program. No comment letters were received by WS within the said comment period.

## Affected Environment

The areas of the proposed action include the entire State of Florida, but more specifically, areas where predation losses to listed species has occurred or may occur in the future. The proposed action could occur on private or public properties within the State of Florida.

## Objectives

The objectives of the proposed action are to:

1) Respond to 100% of the requests for assistance with the appropriate action (technical assistance or direct control) as determined by Florida WS personnel, applying the ADC Decision Model (Slate et al. 1992).

2) Hold sea turtle nest predation to less than 20% per year, on properties with a federal WS operational program.

3) Hold American crocodile nest predation to less than 20% per year, on properties with a federal WS operational program.

4) Hold beach mouse and nesting-wintering shorebird predation to less than 20% per year, on properties with a federal WS operational program.

5) Reduce feral hog populations to the greatest extent possible, on properties with a federal WS operational program.

6) Maintain the lethal take of nontarget animals by WS personnel during damage management to less than 10% of the total animals taken.

## Major Issues

Several major issues were contained in scope of this EA. These issues were consolidated into the following 6 primary issues to be considered in detail:

1) Effects of Predation on Resources Protected, Including Native Wildlife and Plant Species

2) Effects on Target Species Populations

3) Effects of Control Methods on Nontarget Species Populations, Including T&E Species

4) Humaneness of Control Methods

5) Effects of Control Methods on Human Health and Safety

6) Effects on the Aesthetic Values of Targeted Species and Protected T&E Species

**Alternatives Analyzed in Detail**
Five potential alternatives were developed to address the issues identified above. A detailed discussion of the anticipated effects of the alternatives on the objectives and issues are contained in the EA. The following summary provides a brief description of each alternative and its anticipated impacts.

**Alternative 1 - No Action** - This alternative precludes any and all WDM activities by WS to protect T&E species in Florida. A natural resource manager or any other entity directed at preventing or reducing predation of sea turtle nests, crocodile nests, beach mice, and shorebirds could conduct WDM activities in the absence of WS involvement.

**Alternative 2 - Nonlethal Control Before Lethal Control** - This alternative would not allow the use or recommendation of lethal control by WS until all available nonlethal methods had been applied and determined to be inadequate in each damage situation.

**Alternative 3 - Nonlethal Control Only** - This alternative would involve the use and recommendation of nonlethal management techniques only by WS.

**Alternative 4 - Lethal Control Only** - This alternative would involve the use and recommendation of lethal management techniques only by WS.

**Alternative 5 - Integrated Wildlife Damage Management (the Proposed Action)** - This alternative would incorporate an integrated approach to wildlife damage management using components of the wildlife damage management techniques and methods addressed in Alternatives 2, 3, and 4, as deemed appropriate by WS and other participating entities.

**Alternatives Considered but not Analyzed in Detail with Rationale**

1) Aversive Conditioning (taste aversion) Alternative -The objective of aversive conditioning would be to feed egg predators a prey-like bait (eggs) laced with an aversive agent that causes them to become ill, resulting in the subsequent avoidance of the prey (eggs).

The use of any taste aversive agent would be experimental. No compounds are currently registered by the Environmental Protection Agency (EPA) for use in this situation. While some aversive conditioning studies involving raccoons and ravens have proven successful, results with coyotes, wild hogs, and armadillos have been less conclusive. To be successful the predator must be enticed to eat the egg baits; the predator aversive agent used must induce enough discomfort to condition the predator to avoid the baits; and this avoidance must be transferred to sea turtle and shorebird nests. Furthermore, the avoidance must persist long enough without reinforcement for this method to offer realistic protection to sea turtle, crocodile, and shorebird eggs. This method would not address the problem with predation on beach mice, shorebirds, nor sea turtle and crocodile hatchlings.

2) Frightening Devices Alternative - Frightening devices such as electronic guards, pyrotechnics, propane cannons, and lights can be used to temporarily alleviate predation. The effectiveness of these devices depends upon the individual predator's fear of, and subsequent aversion to the offensive stimuli. Once a predator habituates to these stimuli, it often resumes its normal activities and movements.

The continuous and prolonged utilization of artificial lighting along the beach could have a negative impact on sea turtle, crocodile, and shorebird nesting activity, and endangered beach mice foraging. The use of artificial lighting may deter female sea turtles (Witherington and Martin 1996) and shorebirds, discouraging them from nesting at historic nesting sites. In addition, newly hatched sea turtles are strongly attracted to light sources (Raymond 1984, Witherington 1995, Witherington 1991). This disorientation could lead to increased mortality due to predation, dehydration, and exhaustion. Lights could inhibit the foraging behavior of beach mice, since they forage during nighttime hours.

The impact of noise resulting from the use of electronic guards, pyrotechnics, and propane exploders in sea turtle and crocodile nesting areas is unknown. There are indications that the noise and harassment associated with increasing boat and jet ski traffic may stress sea turtles that are feeding, mating, or waiting to nest near popular beaches. Noise associated with the above devices, potentially could impact all animal species proposed for protection in this EA.

After consultation with the FPS and the USFWS, it was decided that this method was unacceptable for use during the sea turtle nesting season (May 1 to October 31), because of the potential impacts to adult nesting and hatchling sea turtles. This method could be used outside of the turtle nesting season from November 1 to April 30; however, the foraging activities of the beach mouse and wintering shorebirds would still be effected by the lights and noise from the frightening devices during this period. Also, using frightening devices during this time would not prevent predation of sea turtle and shorebird nests during nesting season.

Due to the public nature of the Florida coastal environs, and the presence of overnight campers, the use of electronic guards, pyrotechnics, and propane exploders would negatively impact the serene environment. The exclusive use of frightening devices in a manner compatible with park management and sea turtle nesting requirements would not reduce predation to an acceptable level.

3) Population Reduction (trap/translocate) Alternative - This alternative would allow the live capture of raccoons, foxes, coyotes, feral /free-ranging domestic cats and dogs, feral hogs, and armadillos using cage traps, leg snares, and/or leghold traps. Captured predators would be tranquilized and translocated to other areas.

The FWC, Title 39-4.005 (*Introduction of Foreign Wildlife or Freshwater Fish or Carriers of Disease*) does not allow the transportation of non-indigenous wildlife into or within the State of

Florida. For the scope of this EA, this includes feral hogs, cats, dogs, and coyotes. Additionally, relocation of live furbearers (i.e., raccoons, coyotes, foxes, opossums, skunks, nutria, beaver) or nonprotected wildlife (i.e., armadillos) is not permitted in Florida without a permit issued by the FFWCC (FWC, Title 39-24.002 and 39-6.002).

Relocation of wildlife is often viewed as inhumane and biologically unsound management, especially when the wildlife species being relocated is already abundant or common in an area. Relocated animals are forced into a new environment where they often have to compete for space and resources with already well established animals of the same species. Consequently, WS will not request a permit from the state in regards to relocating any of the species proposed for control work in this EA. If certain segments of the public demand relocation, then it will be up to that group(s) to acquire a permit from the state and relocate the animals (as outlined in the relocation permit).

4) Eradication and Long Term Population Suppression of Native Wildlife Alternative - Eradication and long term population suppression of native wildlife is not an objective or option considered by the Wildlife Services Program in Florida. Eradication of native wildlife populations or species is considered ecologically unsound by the Wildlife Services Program, and is not and will not be conducted by WS. Within the scope of this EA, it is the objective of WS to reduce predator numbers within local populations that are directly impacting state and/or federally listed species. However, this reduction will be restricted to problem animals, species, or populations, and will only be conducted with non-native problem species and non-listed native carnivores/omnivores that have been identified as significant predators of listed species in this EA. Additionally, non-native species (i.e., feral hogs) that directly impact the habitats of the listed species will be managed to reduce habitat degradation in these areas and to reduce their impact on other sensitive native fauna, flora, and ecosystems.

5) Biological Control Alternative - Biological control is most commonly used to control select evasive plant and insect species. Very little effort has been devoted to the biological control of wildlife species listed in this EA for two reasons: 1) many of these species are native to the North American continent and biological control measures directed towards a wide spread species potential could have disastrous, uncontrollable effects on a species throughout its range and 2) any biological control measure directed towards a non-native or feral species could adversely affect some groups of animals presently in use for agriculture purposes, ranching, pets, etc. that are closely related to the target species.

**Finding of No Significant Impact (FONSI)**
The analysis in the EA indicates that there will not be a significant impact, individually or cumulatively, on the quality of the human environment as a result of this proposed action. I agree with this conclusion and, therefore, find that an EIS need not be prepared. This determination is based on the following factors:

1)   Predator damage management, as conducted by WS in the State of Florida, is not regional or national in scope.

2)   Based on the analysis documented in the EA, the impacts of the proposed action will not significantly affect public health or safety. Risks to the public from WS methods were determined to be low in a formal risk assessment (USDA 1997, Appendix P).

3)   The proposed action will not have a significant impact on unique characteristics such as park lands, wetlands, wild and scenic areas, or ecologically critical areas. Built-in mitigation measures that are part of WS's standard operating procedures and adherence to laws and regulations will further ensure that WS activities do not harm the environment.

4)   The effects on the quality of the human environment are not highly controversial. Although certain individuals may be opposed to managing predators, this action is not controversial in relation to size, nature, or effects.

5)   Mitigation measures adopted and/or described as part of the proposed action minimize risks to the public, prevent adverse effects on the human environment, and reduce uncertainty and risks. Effects of methods and activities, as proposed, are known and do not involve uncertain or unique risks.

6)   The proposed action does not establish a precedent for future actions, including future predator damage management that may be implemented or planned within the State.

7)   The number of predators that will be taken by WS annually is very small in comparison to regional and statewide populations. Adverse effects on other wildlife species and on wildlife habitat would be minimal. The EA discussed cumulative effects of WS on target and non-target species populations and concluded that such impacts were not significant for this or other anticipated actions to be implemented or planned within the State.

8)   This action will not adversely affect districts, sites, highways, structures, or objects listed in or eligible for listing in the National Register of Historic Places and will not cause loss or destruction of significant scientific, cultural, or historic resources. Wildlife damage management would not disturb soils or any structures and, therefore, would not be considered a "Federal undertaking" as defined by the National Historic Preservation Act.

9)   WS determined that the proposed project would not adversely affect Federally or State listed species in Florida.

10)  The proposed action is consistent with local, state, and Federal laws that provide for or restrict WS wildlife damage management. Therefore, WS concludes that this project is in compliance with federal, state and local laws for environmental protection.

**Decision and Rational**

I have carefully reviewed the Environmental Assessment (EA) prepared for this proposal and the input from the public involvement process. I believe that the issues identified in the EA are best addressed by selecting Alternative 5 (*Integrated Wildlife Damage Management - Proposed Action*) and applying the associated mitigation measures discussed in Chapter 3 of the EA. Alternative 5 is selected because (1) it offers the greatest chance at maximizing effectiveness and benefits to resource owners and managers while minimizing cumulative impacts on the quality of the human environment that might result from the program's effect on target and non-target species populations; (2) it presents the greatest chance of maximizing net benefits while minimizing adverse impacts to public health and safety; and, (3) it offers a balanced approach to the issues of humaneness and aesthetics when all facets of these issues are considered. The comments identified from public involvement were minor and did not change the analysis. Therefore, it is my decision to implement the proposed action as described in the EA.

Copies of the EA are available upon request from the USDA, APHIS, WS, 2820 East University Avenue, Gainesville, FL 32641.


_____                              _1/18/02_____
Acting Director, Eastern Region                          Date
USDA-APHIS-WS

## Literature Cited

Raymond, P. W. 1984. Sea Turtle hatchling disorientation and beach front lighting. A review of the problem and potential solutions. The Center for Environmental Education. Sea Turtle Rescue Fund. Washington, D. C. 72 pp.

Slate, D. A., R. Owens, G. Connolly, and G. Simmons. 1992. Decision making for wildlife damage management. Trans. North Am. Wildl. Nat. Resour. Conf. 57: 51-62.

USDA. 1997. Final Environmental Impact Statement. U.S. Dept. Agric., Anim. Plant Health Inspection Serv., Animal Damage Control, Operational Support Staff, 4700 River Road, Unit 87, Riverdale, MD 20737.

Witherington, B. E. 1991. Influences of artificial lighting on the seaward orientation of hatchling loggerhead turtles. *(Caretta caretta)*. Biological Conservation 55:139-149.

_____. 1995. Hatchling orientation. A Summary. Pages 577-578 in K. A. Bjorndal, ed., Biology and Conservation of sea turtles. Revised Edition. Smithsonian Institution Press. Washington D.C.

_____ and R. E. Martin. 1996. Understanding, assessing, and resolving light-pollution problems on sea turtle nesting beaches. FMRI Tech. Rep. TR-2. Florida Marine Research Institute, St. Petersburg, Florida. 73p.

**Appendix E. Environmental Analysis Documentation for Enhanced Management of Avian Breeding Habitat Injured by Response in the Florida Panhandle, Alabama, and Mississippi**

# ENVIRONMENTAL ANALYSIS
### Supporting Supplemental Environmental Assessment for an Enhanced Management of Avian Breeding Habitat Injured by Response Activities in the Florida Panhandle and on Department of the Interior Lands in Alabama and Mississippi

### Prepared by
### United States Department of the Interior
### United States Fish and Wildlife Service
### National Park Service

**Proposed Action:** Implementation of a coordinated program for enhanced management of avian breeding habitat injured by response in the Florida Panhandle and on the United States Department of the Interior (DOI) lands in Alabama and Mississippi. See Section 4.3 of the Draft Phase II Early Restoration Plan and Environmental Review for the background, purpose and need, and scope of the Proposed Action.

**No Action Alternative:** Under the No Action Alternative, the Trustees would not implement the Proposed Action and would rely solely on natural recovery to restore natural resources and associated services until the natural resource damage assessment and final restoration are complete. Choosing the No Action Alternative, at this time, would not preclude analysis and implementation of different restoration activities at a later date.

**Affected Environment:** See Section 3.3.2.1.1 of the Draft Phase II Early Restoration Plan and Environmental Review.

**Pre-existing Environmental Analysis Adopted by the United States Fish and Wildlife Service (FWS) and the National Park Service (NPS) and Incorporated by Reference:** DOI regulations for implementing the National Environmental Policy Act (NEPA) provide that a DOI bureau may adopt an Environmental Assessment (EA) prepared by another agency [see 43 Code of Federal Regulations (C.F.R.) 46.320]. For the Proposed Action, FWS and NPS have adopted the USDA-Wildlife Services (WS) EA entitled *"Environmental Assessment and Finding of No Significant Impact for Management of Predation Losses to State and Federally Endangered, Threatened, and Species of Special Concern; and Feral Hog Management to Protect Other State and Federally Endangered, Threatened, Species of Special Concern, and Candidate Species of Fauna and Flora in the State of Florida"* (see Appendix D).

**Additional Environmental Analysis Included in this Draft Supplemental EA:** The DOI regulations also provide that, when a bureau's proposed action differs from the proposed action contained in the adopted EA, the bureau may augment the adopted EA to make it consistent with the bureau's proposed action (see 43 C.F.R. 46.320). This Draft Supplemental EA augments the WS EA. In addition to the environmental analysis regarding predator control activities contained in the adopted WS EA, this Draft Supplemental EA considers any additional environmental impacts that would result from the elements of the Proposed Action (i.e., symbolic fencing and signage, and increased surveillance, outreach, and training activities) that are not described and analyzed in the adopted WS EA.

**Environmental Consequences of the Proposed Action:** The Trustees have concluded that the Proposed Action would not result in a significant impact, individually or cumulatively, on the quality of the human environment. On balance, the Proposed Action would have positive effects that are consistent with long-term planning goals and contribute beneficially to avian breeding habitat in Florida and on DOI lands in Alabama and Mississippi. Additionally, all effects are local to the project areas, geographically disparate, and are not expected to overlap the activities or locations of other early restoration projects.

The following table summarizes the WS EA and the FWS and NPS analysis of potential effects from implementing the Proposed Action.

**Environmental Consequences of the No Action Alternative:** The No Action Alternative is used in this analysis a basis for comparison of the effects from implementing the alternatives. The baseline for comparison of the alternatives is defined as the current condition and expected future condition in the absence of the proposed action. Therefore, if the Proposed Action is not implemented (No Action), the injury associated with disturbance of the nesting habitat resulting from the response will be left to natural recovery processes only.

| Issue Analyzed | Short-term Impacts | Long-term Impacts | Indirect Impacts | Cumulative Impacts | Rationale |
|---|---|---|---|---|---|
| Geological resources | No | No | No | No | The proposed project has no potential to affect geological resources. |
| Air quality | No | No | No | No | The proposed project has no potential to affect air quality. |
| Water quality | No | No | No | No | The proposed project has no potential to affect water quality. |
| Soundscapes | No | No | No | No | The proposed project has no potential to affect soundscapes. |
| Marine and estuarine resources | No | No | No | No | The proposed project has no potential to affect marine or estuarine resources. |
| Wetlands and floodplains | No | No | No | No | The proposed project has no potential to affect wetlands or floodplains. |
| Threatened and Endangered Species | Beneficial for beach nesting shorebirds | Beneficial for beach nesting shorebirds | Beneficial for beach nesting shorebirds | Beneficial for beach nesting shorebirds | The adopted WS EA concluded, with concurrence from FWS, that predator control activities "would not likely adversely affect any species protected under the U.S. Endangered Species Act." Consultation under Section 7 of the Endangered Species Act for the entire proposed project would be completed prior to project implementation. The proposed project would be implemented in accordance with all applicable laws and regulations concerning the protection of threatened and endangered species and their habitats. The Trustees are proposing this project because they believe that predator control, symbolic fencing and signage, and increased surveillance, training, outreach, and monitoring activities would have a beneficial impact on the nesting habitat to support the breeding success of beach nesting shorebirds. |

| Issue Analyzed | Short-term Impacts | Long-term Impacts | Indirect Impacts | Cumulative Impacts | Rationale |
|---|---|---|---|---|---|
| Other wildlife and wildlife habitat | The WS EA addresses the effects of predator control. The additional proposed project activities may have minor, short-term and localized effects on other wildlife and habitat. | No | The WS EA addresses the effects of predator control. The additional proposed project activities may have minor, short-term and localized effects on other wildlife and habitat. | No | The adopted WS EA concluded that the number of predators that would be taken annually is very small in comparison to regional and statewide populations. Adverse effects on other wildlife species and habitat would be minimal. The WS EA evaluated cumulative effects on target and non-target species populations and concluded that such impacts were not significant for this or other anticipated actions to be implemented or planned within the State. The Trustees have determined that predator control, symbolic fencing and signage, and increased surveillance, training, outreach, and monitoring activities would not have a significant impact on wildlife in general. |
| Introduce or promote non-native species | No | No | No | No | The proposed project has no potential to introduce or promote the spread of non-native species. |

| Issue Analyzed | Short-term Impacts | Long-term Impacts | Indirect Impacts | Cumulative Impacts | Rationale |
|---|---|---|---|---|---|
| Cultural and historic resources | No | No | No | No | The adopted WS EA concluded that predator control activities would not cause ground disturbances or otherwise have the potential to significantly affect visual, audible, or atmospheric elements of historic properties and are thus not undertakings as defined by the National Historic Preservation Act of 1966 (NHPA). Seasonal symbolic fencing (e.g., driving stakes into the ground and signage) would be done in compliance with the NHPA. Review of the proposed project under Section 106 of the NHPA for the entire proposal would be completed prior to project implementation. The proposed project would be implemented in accordance with all applicable laws and regulations concerning the protection of cultural and historic resources. |
| Other agency or tribal land use plans or policies or private land use | No | No | No | No | The proposed project has no potential to affect other agency or tribal land use plans or policies. The proposed project has no potential to affect private land use. |

| Issue Analyzed | Short-term Impacts | Long-term Impacts | Indirect Impacts | Cumulative Impacts | Rationale |
|---|---|---|---|---|---|
| Socio-economics, minority and low-income populations | No | No | No | No | The adopted WS EA concluded that mitigation measures adopted and/or described as part of predator control activities minimize risks to the public, prevent adverse effects on the human environment, and reduce uncertainty and risks. Effects of predator control methods and activities, as proposed, are known and do not involve uncertain or unique risks. The Trustees have determined that predator control, symbolic fencing and signage, and increased surveillance, training, outreach, and monitoring activities would have no significant effect on socioeconomic or environmental justice issues. |
| Visitor experience and aesthetic resources | Symbolic fencing and signage may have minor, short-term and localized effects on beach aesthetics and visitor experience | No | No | No | The adopted WS EA concluded that predator control activities would not cause significant impacts. In addition, the Trustees have determined that symbolic fencing and signage in place during the nesting season could have minor, short-term and localized impacts on visitor experience and aesthetics during those times when the fences and signage are in place. |

| Issue Analyzed | Short-term Impacts | Long-term Impacts | Indirect Impacts | Cumulative Impacts | Rationale |
|---|---|---|---|---|---|
| Public safety | No | No | No | No | Based on the analysis contained in the adopted WS EA, predator control activities would not significantly affect public health or safety. Risks to the public from WS predator control methods were determined to be low in a formal risk assessment (see WS EA). In addition, the Trustees have determined that symbolic fencing and signage, and increased surveillance, training, outreach, and monitoring activities would have no significant effect on public safety. |
| Energy resources | No | No | No | No | The proposed project has no potential to affect energy resources. |
| Cumulative effects | No | No | No | No | The adopted WS EA concluded that predator control activities would have no significant or cumulative adverse environmental consequences. The Trustees have determined that when combined with past, present and future foreseeable projects, no significant adverse cumulative impacts are anticipated from the proposed project. Additionally, all effects would be local to the project areas, geographically disparate, and are not expected to overlap the activities or locations of other early restoration projects. |

| Issue Analyzed | Short-term Impacts | Long-term Impacts | Indirect Impacts | Cumulative Impacts | Rationale |
|---|---|---|---|---|---|
| Controversial environmental effects | No | No | No | No | The adopted WS EA concluded that the effects on the quality of the human environment from the predator control activities are not highly controversial. Although certain individuals may be opposed to managing predators, the proposed action is not controversial in relation to its size, nature, or effects. In addition, the Trustees have determined that symbolic fencing and signage, and surveillance, training, outreach, and monitoring activities are common management activities and would not be controversial. |
| Establish a precedent for future actions | No | No | No | No | The proposed project would not establish a precedent for future actions. The Trustees have determined that predator control, symbolic fencing and signage, and surveillance, outreach, and monitoring activities are well-established management activities. |

**Appendix F. Environmental Analysis Documentation for Improving Habitat Injured by Spill Response: Restoring the Night Sky**

# UNITED STATES FISH AND WILDLIFE SERVICE
# ENVIRONMENTAL ACTION STATEMENT

Within the spirit and intent of the Council on Environmental Quality's regulations for implementing the National Environmental Policy Act (NEPA), and other statutes, orders, and policies that protect fish and wildlife resources, I have established the following administrative record and determined that the action of "Restoring the Night Sky"(retrofitting existing street lights to reduce visibility from the beach to improve sea turtle nesting habitat injured by spill response within the states of Florida and Alabama) as described in the attached Deepwater Horizon Oil Spill Draft Early Restoration Plan meets the following USFWS resource management categorical exclusions:

- **516 DM 8.5A(2)**-Personnel training, environmental interpretation, public safety efforts, and other educational activities, which do not involve new construction or major additions to existing facilities.
- **516 DM 8.5B(2)**-The operation, maintenance, and management of existing facilities and routine recurring management activities and improvements, including renovations or replacements which result in no or only minor changes in the use, and have no or negligible environmental effects on-site or in the vicinity of the site.
- **516 DM 8.5B(11)**-Natural resource damage assessment restoration plans, prepared under sections 107, 111, and 122(j) of the Comprehensive Environmental Response Compensation and Liability Act (CERCLA); section 311(f)(4) of the Clean Water Act; and the Oil Pollution Act; when only minor or negligible change in the use of the affected areas is planned.

Check One:

__X__ is a categorical exclusion as provided by 516 DM 8.5 A(2), B(2) and B(11). No further NEPA documentation will therefore be made.

_____ is found not to have significant environmental effects as determined by the attached environmental assessment and finding of no significant impact.

_____ is found to have significant effects and, therefore, further consideration of this action will require a notice of intent to be published in the Federal Register announcing the decision to prepare an EIS.

_____ is not approved because of unacceptable environmental damage, or violation of Fish and Wildlife Service mandates, policy, regulations, or procedures.

_____ is an emergency action within the context of 40 CFR 1506.11. Only those actions necessary to control the immediate impacts of the emergency will be taken. Other related actions remain subject to NEPA review.

Other supporting documents (list):
See attached Deepwater Horizon Oil Spill Draft Early Restoration Plan.

Signature Approval:

| (1) Originator Coordinator | Date | (2) WO/RO Environmental | Date |
| (3) AD/ARD | Date | (4) Director/Regional Director | Date 10/24/12 |

03129196 FWM 246
New

ENVIRONMENTAL QUALITY

NEPA COMPLIANCE CHECKLIST

State: FL, AL

Federal Financial Assistance Grant/Agreement/Amendment Number: N/A

Grant/Project Name: Enhancement of Sea Turtle Nesting Habitat "Restore the Night Sky"

This proposal X is; ☐ is not completely covered by categorical exclusion in: 516 DM 8.5A(2), B(2), B(11).

*(check (X) one) (Review proposed activities. An appropriate categorical exclusion must be identified before completing the remainder of the Checklist. If a categorical exclusion cannot be identified, or the proposal cannot meet the qualifying criteria in the categorical exclusion, or an extraordinary circumstance applies (see below), an EA must be prepared.)*

## Extraordinary Circumstances:

Will This Proposal *(check (X) yes or no for each item below):*

| Yes | No | | |
|---|---|---|---|
| ☐ | X | 1. | Have significant adverse effects on public health or safety. |
| ☐ | X | 2. | Have significant adverse effects on such natural resources and unique geographic characteristics as historic or cultural resources; park, recreation or refuge lands; wilderness areas; wild or scenic rivers; national natural landmarks; sole or principal drinking water aquifers; prime farmlands; wetlands (Executive Order 11990); floodplains(Executive Order 11988); national monuments; migratory birds (Executive Order 13186); and other ecologically significant or critical areas under Federal ownership or jurisdiction. |
| ☐ | X | 3. | Have highly controversial environmental effects or involve unresolved conflicts concerning alternative uses of available resources [NEPA Section 102(2)(E)]. |
| ☐ | X | 4. | Have highly uncertain and potentially significant environmental effects or involve unique or unknown environmental risks. |
| ☐ | X | 5. | Have a precedent for future action or represent a decision in principle about future actions with potentially significant environmental effects. |
| ☐ | X | 6. | Have a direct relationship to other actions with individually insignificant but cumulatively significant environmental effects. |
| ☐ | X | 7. | Have significant adverse effects on properties listed or eligible for listing on the National Register of Historic Places as determined by either the bureau or office, the State Historic Preservation Officer, the Tribal Historic Preservation Officer, the Advisory Council on Historic Preservation, or a consulting party under 36 CFR 800. |
| ☐ | X | 8. | Have significant adverse effects on species listed, or proposed to be listed, on the List of Endangered or Threatened Species, or have significant adverse effects on designated Critical Habitat for these species. |
| ☐ | X | 9. | Have the possibility of violating a Federal law, or a State, local, or tribal law or requirement imposed for the protection of the environment. |
| ☐ | X | 10. | Have the possibility for a disproportionately high and adverse effect on low income or minority populations (Executive Order 12898). |
| ☐ | X | 11. | Have the possibility to limit access to and ceremonial use of Indian sacred sites on Federal lands by Indian religious practitioners or significantly adversely affect the physical integrity of such sacred sites (Executive Order 13007). |
| ☐ | X | 12. | Have the possibility to significantly contribute to the introduction, continued existence, or spread of noxious weeds or non-native invasive species known to occur in the area or actions that may promote the introduction, growth, or expansion of the range of such species (Federal Noxious Weed Control Act and Executive Order 13112). |

*(If any of the above extraordinary circumstances receive a "Yes" check (X), an EA must be prepared.)*

X Yes ☐ No   This grant/project includes additional information supporting the Checklist.

## Concurrences/Approvals:

Project Leader: _Robin Renn_                                   Date: 10-23-2012

State Authority Concurrence: _____     Date: _____
*(with financial assistance signature authority, if applicable)*

*Within the spirit and intent of the Council of Environmental Quality's regulations for implementing the National Environmental Policy Act (NEPA) and other statutes, orders, and policies that protect fish and wildlife resources, I have established the following administrative record and have determined that the grant/agreement/amendment:*

X      is a categorical exclusion as provided by 516 DM 8.5 and/or 43 C.F.R. 46.210. No further NEPA documentation will therefore be made.

        is not completely covered by the categorical exclusion as provided by 516 DM 8.5 and/or 43 C.F.R. 46.210. An EA must be prepared.

## Service signature approval:

RO or WO Environmental Coordinator: _Cynthia K. Dohner_   Date: 10/24/2012

Staff Specialist, Division of Federal Assistance: _____   Date: _____
*(or authorized Service representative with financial assistance signature authority)*

FWS Form 3-2185

**Appendix G. Finding of No Significant Impact for Enhanced Management of Avian Breeding Habitat Injured by Response in the Florida Panhandle, Alabama, and Mississippi**

# FINDING OF NO SIGNIFICANT IMPACT

***Enhanced Management of Avian Breeding Habitat Injured by Response Activities in the Florida Panhandle and on Department of the Interior Lands in Alabama and Mississippi***

State and federal natural resource trustees for the *Deepwater Horizon* Oil Spill Natural Resource Damage Assessment and Restoration (NRDAR) (collectively, the Trustees), propose to implement the early restoration project entitled *Enhanced Management of Avian Breeding Habitat Injured by Response Activities in the Florida Panhandle and on Department of the Interior Lands in Alabama and Mississippi* (Proposed Action).

The Proposed Action involves three categories of activities:

1) Placing symbolic fencing around sensitive bird nesting sites to indicate the site as off-limits to people, pets, and other sources of disturbance;
2) Increasing surveillance and efficacy of posted nesting sites with increased training, outreach, and monitoring by Florida Fish and Wildlife Conservation Commission (FWC), Florida Department of Environmental Protection (FDEP), NPS, and FWS biologists and staff to minimize disturbance to beach nesting birds in posted areas, and;
3) Increasing predator control to reduce disturbance of eggs, chicks, and adult birds at nesting sites in Florida.

In Florida, the Proposed Action would be implemented in Escambia, Santa Rosa, Okaloosa, Walton, Bay, Gulf, and Franklin Counties. In Alabama, the Proposed Action would be implemented on Bon Secour National Wildlife Refuge in Baldwin and Mobile Counties. In Mississippi, the Proposed Action would be implemented on Gulf Islands National Seashore – Mississippi District.

The Proposed Action is an early restoration project funded as part of the *Deepwater Horizon* NRDAR process in accordance with the "Framework for Early Restoration Addressing Injuries Resulting from the *Deepwater Horizon* Oil Spill." It was one of two projects that the Trustees proposed for implementation by the Trustees in a Draft Phase II Early Restoration Plan and Environmental Review (Phase II DERP/ER), and represents initial steps toward the restoration of natural resources injured by the *Deepwater Horizon* Oil Spill.[1] The Trustees have considered the public comments received on the Phase II DERP/ER and now intend to finalize the selection of the Proposed Action as an early restoration project for implementation.

Under the Oil Pollution Act of 1990 (OPA), damages recovered from parties responsible for natural resource injuries are used to restore, replace, rehabilitate and/or acquire the equivalent of the injured natural resources. *See* 33 U.S.C. § 2706. Actions undertaken by federal trustees to restore natural resources under OPA and other federal laws are subject to the National Environmental Policy Act (NEPA), 42 U.S.C. § 4321 *et seq.*

---

[1] FWS has determined that the other proposed project, entitled *Improving Habitat Injured by Spill Response: Restoring the Night Sky*, is categorically excluded under the National Environmental Policy Act pursuant to FWS categorical exclusions listed in 516 DM 8.5(A)(2), B(2), and B(11). *See* Phase II DERP/ER, Appendix F.

To comply with NEPA, NPS and FWS prepared a draft Supplemental Environmental Assessment (EA) for the Proposed Action and a No Action alternative. One of the three categories of activities included in the Proposed Action, predator control in Florida, was previously analyzed in a Final EA prepared by the U.S. Department of Agriculture (USDA), Animal & Plant Health Inspection Service (APHIS), Wildlife Service (WS), entitled *Environmental Assessment and Finding of No Significant Impact for Management of Predation Losses to State and Federally Endangered, Threatened, and Species of Special Concern; and Federal Hog Management to Protect Other State and Federally Endangered, Threatened, Species of Special Concern, and Candidate Species of Fauna and Flora in the State of Florida* (USDA-APHIS-WS, 2002). FWS was a cooperating agency in the preparation of the USDA-APHIS-WS Final EA. For the Proposed Action, NPS and FWS adopted the USDA-APHIS-WS Final EA and incorporated it in the Phase II DERP/ER as Appendix D.

Additionally, NPS and FWS augmented the analysis contained in the adopted USDA-APHIS-WS Final EA to make it fully consistent with the Proposed Action. Thus, in addition to the environmental consequences of predator control in Florida, the draft Supplemental EA considered any environmental impacts that would result from the two categories of activities in the Proposed Action (*i.e.*, symbolic fencing and signage; and increased surveillance, outreach, and training) that are not analyzed in the adopted USDA-APHIS-WS Final EA.

After considering input from the public during the comment period for the Phase II DERP/ER, the Trustees prepared a Phase II Early Restoration Plan/Environmental Review (Phase II ERP/ER), which includes the Supplemental EA. A summary of comments received and agency response is included in the Phase II ERP/ER.

The Proposed Action was selected by the Trustees for implementation because, when compared to the No Action alternative, it would improve the quality and function of nesting habitat used by Gulf beach nesting birds in the project area. Furthermore, aside from these beneficial effects, the Supplemental EA indicates that the Proposed Action would not have significant adverse effects, individually or cumulatively, on the quality of the human environment.

DETERMINATION

Based on my review and evaluation of the Phase II ERP/ER and Supplemental EA, and public comments received on the Phase II DERP/ER and draft Supplemental EA, I have determined that the Proposed Action is not a major federal action which would significantly affect the quality of the human environment within the meaning of Section 102(2)(C) of NEPA, 42 U.S.C. § 4332(2)(C). This Finding of No Significant Impact (FONSI) accepts, incorporates by reference, and augments the USDA-APHIS-WS Final EA issued in 2002. Accordingly, preparation of an Environmental Impact Statement on the Proposed Action is not required.

RATIONALE

In addition to the reasons stated in the FONSI issued by USDA-APHIS-WS in 2002:
- The Proposed Action has no potential to adversely affect the following resources/issues that were analyzed: geological resources, air quality, water quality, soundscapes, marine

and estuarine resources, wetlands and floodplains, introduction and spread of non-native species, other agency, tribal or private land use plans or policies, or energy resources.

- Under the Proposed Action, adverse effects on other wildlife species and habitat would be minimal. The number of predators that would be taken annually is very small in comparison to regional and statewide populations.

- A complete review of the Proposed Action under Section 106 of the National Historic Preservation Act would be conducted prior to project implementation.

- Endangered Species Act Section 7 consultation would be completed prior to implementation of the Proposed Action. The long term effects of improved beach nesting bird habitat are expected to benefit many threatened and endangered bird species.

- Florida, Alabama, and Mississippi determined that the Proposed Action would be consistent with their respective Coastal Zone Management Plans.

- The Proposed Action would have no significant effect on socioeconomic or environmental justice issues.

- The Proposed Action would not cause significant effects to visitor experience and aesthetic resources. Symbolic fencing and signage in place during the nesting season could have minor, short-term and localized impacts on visitor experience and aesthetics during those times when the fences and signage are in place.

- Implementing the Proposed Action would not be highly controversial. Although certain individuals may be opposed to managing predators, the Proposed Action is not controversial in relation to its size, nature, or effects. The Proposed Action consists of common management activities and would not be controversial or establish a precedent for future actions.

- No significant adverse cumulative impacts are anticipated from implementation of the Proposed Action.

- If the Proposed Action is not implemented (No Action), the negative impacts to beach nesting bird habitat that would be avoided through the Proposed Action would be expected to continue.

- A notice of availability and request for comments on the Phase II DERP/ER was published in the Federal Register on November 6, 2012. *See Deepwater Horizon* Oil Spill Phase II Draft Early Restoration Plan and Environmental Review, 77 Fed. Reg. 66626 (Nov. 6, 2012). Public comments on the Phase II DERP/ER, including the draft Supplemental EA, were solicited from November 6, 2012 through December 10, 2012, and a public meeting was held on November 13, 2012 in Pensacola, Florida. No substantive public comments were received that would necessitate a change in the analysis described in the draft Supplemental EA. The Phase II ERP/ER and Supplemental EA are hereby incorporated by reference.

Date: 12 | 14 | 2012

Signature: *[signature]*

Cynthia K. Dohner
Authorized Official, U.S. Department of the Interior

www.ingramcontent.com/pod-product-compliance
Lightning Source LLC
Chambersburg PA
CBHW080638180526
45168CB00008B/3221